GOING OVERBOARD

GOING OVERBOARD

Lucy Gwin

THE VIKING PRESS   NEW YORK

First published in 1982 by The Viking Press
625 Madison Avenue, New York, N.Y. 10022
Published simultaneously in Canada by
Penguin Books Canada Limited

LIBRARY OF CONGRESS CATALOGING IN PUBLICATION DATA
Gwin, Lucy.
Going overboard.
1. Women—Louisiana—Biography. I. Title.
HQ1438.L8G85   305.4'2'09763   81–16408
ISBN 0–670–34360–9   AACR2

Printed in the United States of America
Set in Video Century Expanded

This book is dedicated to Betty c.

I sit down to make an accounting of my year in the offshore oilfields merchant marine with a certain woman in mind, a woman I never met who signs herself Betty c. On my last night as a sailor I found a message from her on the overhead of my bunk.

Women are rare in the Gulf Coast oilfields. In 1979 males outnumbered females hundreds, maybe thousands, to one; a message from one oilfield woman to another was surprising in itself. So I had reason to take this communication seriously, even though it disturbed me.

Bordered by a rude drawing of oversized male genitals and written in the uncertain hand of a near-illiterate, it said: "I like to fuck anytime. Big Dicks for the Cook. Betty c."

After a moment's thought, I set my rebuttal alongside it: "A woman's place is in the wheelhouse." But my dialogue with Betty c. was not so easily concluded.

Flat on my back on that boat, Betty, lust was maybe the least of my preoccupations. Up until the time I went to work on the boats, I'd thought of myself as a man's woman. Women, I believed, led boring and limited lives. Men were free; I thrived on their company. I'd fallen in love no fewer than twenty-two times in my life. But after a year in the merchant marine I would've traded my entire reproductive apparatus for a chance to do my job in peace.

I was the first woman, or maybe only one of the first—Guinness doesn't keep records in this category—to work as a deckhand on the oilfield supply vessels. The work was hard; the men were harder. They tested me, courted me, competed with me, nearly killed me once or twice. Through it all they insisted they knew what women like me were about: "No woman comes out here in the man's world 'less she just wants to get fucked." I denied that, Betty, in the name of female sailors everywhere. But then I found your mark.

Betty, I only wish I knew you, that we could talk. Failing that, I wish you could read. Because today I sit down to write you a book.

Contents

Part One

THE ONLIEST LITTLE WOMAN IN THE OFFSHORE OILFIELDS

1

This is it, the jumping-off point, edge of the known world. Many who come here to make their fortunes, or, like me, only to jump off the world for a while, are not seen again. You may never have heard of Morgan City, Louisiana, but it is a major capital of American blue-collar nomadic culture and hub of the Gulf Coast offshore oilfields. A sign on its outskirts identifies it as "Morgan City—Home of the Shrimp and Petroleum Festival." "One Hundred Years of Progress," reads another.

I would erect a different sign: "Morgan City—Epicenter of Too Many Booms. One Hundred Years of Boomism."

In 1874 Morgan City built a moss factory and enjoyed a modest boom in sphagnum harvested from the surrounding Atchafalaya Basin swamp. The arrival of the railroads brought full-scale booms in timber and shipbuilding, a boomlet in beeswax and honey, booms in pelican oil, otter and beaver pelts, egret plumes, alligator hides. Then came the booms in crab meat, oysters, pogyfish, jumbo shrimp. Never mind that the Pelican State no longer has a pelican to its name —nor an otter, nor a beaver—or that its alligator and egret populations were saved from extinction only by unwelcome federal intervention, or that the oyster and shrimp catches get thinner every year; a boom is a boom, irresistibly here today while the getting's good, gone tomorrow when the boomstuff runs out. Morgan City stands in the ruins of her sequential booms and seems not to mind the temporariness of it all. The booms just keep coming.

In 1949 came the most powerful economic explosion to date: the boom in black gold, crude oil. If the oil lay beneath Morgan City's main street, the city fathers would surely have called in the bulldozers and wiped their town off the map. But the oil is pooled downstream and offshore, beneath the blue gumbo mud and coral shelves of the Gulf of Mexico. Morgan City, with its well-developed port and its central location, is the oilfields' major freight and manpower depot.

This boom too shall pass. The oil companies are pumping it dry as fast as they can pipe the oil to their refineries. But Morgan City's boosters speak with confidence of the coming boom in sea-griculture. The indigenous Cajun population puts it this way: "Don't worry about nothin'."

These Louisiana Cajuns are descendants of peasant French d.p.'s who were driven out of their settlements in Canada, subsequently driven out of the West Indies, driven finally to the yellow fever swamps of the Gulf Coast—where they in turn drove out the Attakappas Indians. Their history bears at least a superficial up-against-the-wall resemblance to the Israelis'. The Cajuns are as entrenched and nearly as beleaguered. The alien invaders are Texans, Floridians, Alabamians, Mississippians, Vietnamese—all come for a share of the oilfield plenty, all men without a country now.

I arrived on the oilfield coast unprepared for its twentieth-century frontier bleakness. My own home had been a well-mannered white-collar city thirteen hundred miles to the northeast. Morgan City was as foreign to my sensibilities as Easter Island; another country, not my country at all. If I'd known Morgan City then the way I know it now, I would've picked another destination. But I hadn't thought it over in advance. I was running away, and down to my last twenty dollars.

As my tired old truck labored up onto the crest of the Bayou

Boeuf Bridge, I caught a first panoramic view of Morgan City. There it was, squatting flat and unlovely over the mouth of the Atchafalaya River, crowded into the bare and narrow ground between Louisiana Highway 90 and an endless jungly swamp. At the foot of the bridge a burnt-out barroom sign spoke in Cajun quasi-French: "Le Bon Ton Rool-ay" (Tr.: Let the good times roll). A five-mile strip of pipeyards followed. Pipeyards and drydocks, helicopter pads and franchiseburger stores, trailer parks, tattoo parlors, and handgun supermarkets. It appeared that some monster wave had rolled over Morgan City months before, that only now a gerry-rigged city of trailers and hasty-pasty aluminum and pole constructs was rising from the ruins. On closer inspection I began to see that the piecemeal look of things might have resulted from oilfield boomism and not from natural disaster. I saw, too, that this was a city of men; men working, or anyway in work clothes, jumpsuited, baseball-capped, bearded, rough, and apparently ready. Everywhere I looked, men looked back at me. Where were all the women? Where were all the homes?

But Morgan City is no American dreamtown of homes and churches. It's a manpower town, a freight dock, a staging area. Around the clock the oilfield supply ships arrive at the docks, load up with supplies for the rigs, and run back down the Atchafalaya to the Gulf. Because offshore oil rigs are man-made islands, the supply vessels—and to some extent, a fleet of helicopters—must ferry the rigs' groceries, their fuel and water, their machinery and men. The men who crew the rigs and the supply ships work in shifts of one or more weeks at a time without a shore break. On the midweek days when the rig and boat crews change over to allow the offshore workers their home leave, Morgan City's single main highway breaks out in clumps of bad-toothed hitchhikers carrying makeshift seabags. (All along the oilfield coast a plastic garbage bag is referred to as a Morgan City Suitcase.) Those lonesome, horny, homeless, hard-muscled men are known to the locals as oilfield trash, *rigrats.*

All rigrats are male. All but a few of them are under thirty years old, Caucasian, nomadic. They lead a bruising life on the underside of American protection and plenty, bumming their way from the Gulf Coast oilfields to the Baltimore Canyon to the Alaska pipeline and back, always back. Some of them get as far away as the oilfields of Africa, Venezuela, the North and China seas. But Morgan City is their training ground, their jumping-off point, the one town they can count on for a ready job if all else fails. Such men are always in demand in the oilfields; the oil companies can never seem to hire on enough of them. Morgan City couldn't prosper without them. But nobody really wants them at all. They are rogue males in the grip of testosterone wanderlust and you can smell it on them.

Out on the rigs and the boats, isolated for weeks at a time from civilization and the comforts of land, they work very nearly around the clock at the dirty, dangerous work that keeps the black gold coming. For this they are paid what must seem to them a great deal of money: in 1978 a minimum of eight dollars an hour for jobs that require no particular skill but brute force. Skilled and especially dangerous work pays more. An underwater welder, for instance, was earning thirty-six dollars an hour in 1978 for simply standing at the ready, three times as much when, as he puts it, he "gets wet." These basic wages are multiplied by guaranteed overtime and double overtime into lurid amounts of boomtown cash.

When the rigrats hit Morgan City to collect their shore leave, some few of them head to their faraway homes. The rest spend their four-figure paychecks with a resolute frenzy, flushing great waves of cash into overpriced motel rooms, heavy-duty armament, silver-toed alligator-skin cowboy boots, solid gold drill bit pendants, and week-long long weekends of hard boozing. Before their seven leave days are over, those same sorry men are hitting up the pawn brokers, sleeping in the bus station waiting room.

Where are all the women, they want to know. If they could find themselves some good women, they say, they'd probably settle down, make a home. Daughters of respectable local families know better; a rigrat is seldom the material from which a good provider can be made. What few local bad girls run with rigrats soon see the light and split town, or turn pro. Morgan City is a man's world. Morgan City can hurt you. Maybe fifty barrooms, from dub-l-wide trailer cocktail lounges to shantytown front porch tavern shacks, collect swarms of Harley hawgs around their front doors. Inside, rigrats and oilfield sailors trade comradely lies and hard-eyed stares, drinking their way to stubborn, insensible fury. At midnight or thereabouts, the accumulated menace is likely to strike sparks. Soon rigrats are howling off in ambulances to St. Mary's Parish Hospital, in paddy wagons to St. Mary's Parish Jail. "They were just born rotten," I heard one deputy complain.

On my first morning in Morgan City I didn't like its looks one bit until I found a place on the docks where I could park my truck and watch the water traffic. I'd been land-locked all my life; this view was fascinating. I saw tugboats pushing chains of barges upriver to the locks with tiny oyster fishers threading their ways between them. An occasional seaplane buzzed on the surface of the bronze river, getting up speed. Supply ships backed into docks where cranes lifted swaying pallets of chemicals and slingloads of steel pipe to their decks. At rotting piers, shrimp boats sat low in the water with their booms at vertical and their bright colors peeling. I liked the shrimpers' romantic-industrial names: *Miss Shirley Girl, Oliphant Guidry, Star Trek,* Sacred Heart Shrimp Company's *Boat #5.* I liked spying out the everyday port life, too: sailors hosing down their decks and calling out to old mates who leaned at the bulwarks of ships passing by; captains up in wheel houses tuning their radar sets and frowning over coastal charts; skinny-legged tall-booted dock boys squatting on metal ramps passing jays to one another, then jumping up when their names were called on the dock's p.a.

Oversized speed yachts barreled up the channel cutting deep wakes; the wakes kicked smaller boats against the docks, causing flag lines to clang against steel masts; it was as if a dozen school bells rang in sequence, echoing. Sea gulls wheeled in the hot morning sky. I told myself everything would work out fine. Morgan City was maybe not as bad as it looked. It was anyway only a dock I'd be stepping over on my way to a job on a boat. A job shouldn't be hard to come by in Morgan City. A highway billboard advertised for "1700 workers wanted NOW at Avondale Shipyards." The local radio station broadcast a stream of employment come-ons interrupted only briefly by hillbilly hero tunes and news flashes from the police blotter. Five of the twelve pages of a Morgan City daily paper were given over to Help Wanteds.

An old friend of mine, Slammin' Sam Baxter, had steered me to Morgan City and a cooking job on the oilfield supply boats. He'd guaranteed I'd get the job I wanted on my first day in town. He'd even furnished me with a map that would lead me to Watercraft, a major oilfield boat company he thought would be hiring at this time of year. I didn't bother with the Help Wanteds; I headed straight for Watercraft. A uniformed guard at the company gate took my name, handed me an application form, and listed me in line for an employment office pass. I waited my turn with twenty other oilfield vagrants in a sheetmetal hiring shed.

It was early in the morning and late in October, but the delta sun bore down out of a cloudless enamel-blue sky, irradiating the shed roof, steaming away our weak attempts at small talk. We barely breathed, and moved only to wipe our sweating faces with our sleeves.

After a while a one-eyed fat man with his change of clothes tied up in a soiled bundle bummed a cigarette from me and struck up a monologue. He said he'd come from Pasadena, Texas, and a lost job as a short-order cook. "I've been up and I've been down, far as this Gulf goes. You b'lieve I used to be a captain? Eyes give out, couldn't pass the test no more. Too old for deckhand. Reckon I'll cook."

His traveling companion, a hard-snoring, coarse-whiskered old boozer, was dressed in a crusty corduroy jumpsuit held together with a plastic duck diaper pin. Twice he pitched forward on his face before we shored him up in a corner. His buddy dusted him off tenderly, sighing.

A pair of hyena-faced young brothers from north Alabama licked the sweat from their upper lips and hooded their eyes, watching but not looking at me. One said he'd been fired from Watercraft last year when his captain caught him smoking dope, but he thought they'd hire him back for deckhand. (They did.)

"When they're hirin', they're hirin'," he told us. "Lookin' for twenty-seven cooks, sixty deckhands today, guard said." Why so many? "Comin' into winter. Gulf gets rough in winter."

Among the dozens of men, one other woman there, a self-confessed whore who told me she'd caught "a bad case of Asia Clap" and needed a rest. "Penicillin don't touch it," she told me, not without a trace of pride. She hoped they'd hire her without a blood test. (They didn't.)

I saw I'd miscalculated my wardrobe. Here I was wearing a pair of shiny new forty-two-dollar deck shoes and a designer jeans pantsuit, carrying a leather briefcase of sample menus and letters of reference. I'd pass the blood test, but I just didn't fit the scene.

I tore up my first try at the application, the one where I admitted to being former vice president of a Michigan Avenue advertising agency, former owner of a gourmet restaurant. I was no Yankee spy, no union organizer or investigative reporter. My running away to sea was as serious to me as the Alabama boy's was to him. I was as broke as he was, too. But my background might raise suspicion. On a fresh application form I demoted myself to former cook and waitress.

I needn't have troubled. Five hours in the steamy hiring shed must have done the trick. Watercraft's hiring secretary put me on without a second look. "You kin cook? Thass what we want. Start Tuesday, week from today. Be at the gate at oh-four-hunnert hours. Bring an alarm clock, no liquor, no mind-altering drugs. All you gotta do is what the cap'n tells you. Cook job pays forty-five dollars a day."

I boiled over with questions. What was the name of my boat? How many in the crew? Where would we be going?

Mainly, what's it like out there?

She answered only the last, wrinkling her little white nose. "I don' have the least idea. I wouldn' go on one of them dirty boats if you paid me."

On my way out of town, I saw a homemade sign standing out in front of a gas station. The sign said: "Fight the Skum People. 4 Armed Robberys here This Year, Criminals not caught Yet."

I winced, and glued my eyes to the road again, walking a tightrope in my getaway from Morgan City. Just balance that unwarranted optimism overhead and keep your eyes front. Don't look down.

I'd arrived before dawn. I was leaving at sunset, heading back to the pinewoods cabin my lover and I had rented the week before, one hundred and fifty miles away above Lake Pontchartrain. Slammin' Sam had warned us to find a home as far from Morgan City as possible. Sam had been right about everything so far. I hoped he'd been just as right about matching me up with the boats. It wasn't as if I had much sea time under my belt. I'd ridden the Staten Island Ferry and the tour boat at Niagara Falls. And just the year before my trip to Morgan City, when the restaurant I owned catered a businessman's cruise on Lake Ontario, I'd cooked aboard a yacht for three days. That last experience had awakened my mild ambition to be a sailor.

The cruise started with a steady rain that I didn't have time to notice; I was busy belowdecks, preparing a gourmet Mexican breakfast. But half an hour out of dock an autumn gale whipped up. The boat lurched, rising and falling arhythmically on graygreen swells. Our passengers and my own hired hands stumbled to the rail and spewed out their breakfasts. Of the twenty-five souls aboard, only the captain and I stayed on our feet, rolling with the boat that plunged in the gale.

Here was one glory of the sea for which Conrad and Melville had not prepared me: the fascist ecstasy of thrill-riding a storm while the poor mortals aboard whine helplessly and puke in their shoes.

One board member vomited into the bowl of sculptured fruit salad I was clearing away, and still my stomach held its own. "This way to the rail, sailor," I told him, guiding him by his well-tailored elbow. Ah, the sea! Ah, stalwart me. A born sailor, the captain said. He said I was such an obvious natural for the sea that I should make a career of sailing. I even considered it, in daydreams. But I had a daughter at home and a restaurant to run. Eight months later the restaurant failed.

My friends all asked me, "What are you going to *do?*" Bankruptcy is like divorce. It brings out the worst in one's friends, who are fighting their own anxious battles. They project their anxiety attacks onto the loser, and furiously too, a behavior that among birds is called mobbing. "What are you going to *do?*"

Damned if I knew. Hadn't I done enough? The way I looked at it, I'd been living out my own and everyone else's fantasy lives too long. There'd been the modest Hoosier married-and-happily-ever-after fantasy that I went for at the age of seventeen and ran out on four years later. A quick succession of sideline fantasies followed: zookeeping, TV station girl fridaying, hotshot legal secretarying. In one long-term fantasy-come-true I put in seven years as an advertising writer, dropping out once to be a hippie, then again to live out a fantasy of homesteading in the farmland of Wisconsin. I turned up my nose at the idea of a real career, of lifelong security. That was for the so-called Realists.

I quit advertising completely, finally, and forever on my thirtieth birthday, tossing away a vice presidency, profit sharing, stock, the works. "You only live once," I said. Let my life seep away to the accompaniment of ventilator whine in Chicago high-rise office buildings? Not me. God, I felt superior.

I retired to Rochester, New York, where I planned to become a student of Roshi Phillip Kapleau, the American Zen master. But after three months of the low-pressure life I'd promised myself, I got itchy for some action. I took an aimless walk one day and wound up signing a lease on an empty storefront. That night I got high with some friends and mapped out the restaurant into which I'd pour my savings. Fantasies can come true. See how easy it is?

Five years later, when the restaurant crashed, I was thirty-five years old and temporarily out of fantasies. Out of money, too. The only thing that seemed easy by then was losing.

My seventeen-year-old daughter, who'd chosen to live with me rather than with her stable father, "out of a sense of adventure," she'd said, had suffered enough poverty and neglect during the restaurant years to keep a someday analyst in a year's supply of fifty-minute hours. She assumed I'd retreat to the security of the ad business. When she found out I had no intention of going back, or any other intention, she defected to her father. I believed I'd lost the right to argue the point; her father was a realist. His life didn't fall apart every few years. Apparently mine did.

He came to get her in his private plane. My younger daughter, who had long been in his keeping, came along for the ride. She seemed more distant from me than ever. I said goodbye to my children, then I crumpled. If I was no longer to be a mother, a restaurant owner, a neighborhood somebody, just who was I to be? What was I for?

I remember asking myself those questions all one month, the month when I sold off all but a few of my possessions to help pay my debts. At the end of that month I had nothing left but a hollow dread of failing again and a lover, whom I shall call Seymour.

Seymour was no prize. A would-be writer, a part-time house painter, a recovering alcoholic . . . all right, all right, he was a bum. But a great talker and a sexual genius. We shared a loathing for lunch boxes and briefcases, health food and organized religion. Both of us loved old roadside motels, marijuana, philosophic speculation, and sex. If I had a major complaint about him, it was that he was too polite, a passive aggressive who believed it was rude to say what he wanted but expected to get it all the same. I spent too much of our time together guessing his wants, tiptoeing around his moods as if they were land mines. Sometimes, especially when he fell off the wagon, they were.

Even so, when I think of him now I picture the two of us sprawling in comfortable chairs with our feet up, smoking dope and writing love letters to each other. We filled one tall basket with those letters. Apparently he loved me above all other women. I certainly loved him. To me, as to a number of good women who had preceded me, Seymour was no pint-sized world-worn loser but a kind of prince of pain and joy. He was also a migratory bird. Winter was coming on. We were north. He wanted to head south. I waffled on that. It was too easy to picture myself stranded, alone in some seedy back-street apartment with Seymour long gone.

Slammin' Sam, a young old friend, cut my indecision out from under me when he came back from a year of adventuring at sea. He spent a long afternoon telling me about his winter on the New England scallop fishing boats, his spring-time job on a genuine brigantine in the Caribbean, his summer as a cook on the Gulf Coast oilfield supply boats. That last job, he said, was made for me. I'd be cooking for five people three times a day with the rest of my time to myself. I'd have an air-conditioned private stateroom and one week off after every week of duty. Sam thought my patches of conversational French would come in handy with the Cajun boat captains. He even thought that the shortage of women in the oilfields would work to my advantage. And just think of the stories I could tell my grandchildren . . .

For the first time since the fall of the restaurant, I felt a goosy thrill of confidence. Maybe Sam had something there. I could turn this thing around and live out the kind of adventure that would arouse not pity but envy. I'd have, at least, an answer to my friends' troubling question. What was I going to do? I was running away to sea. Chew on that, land-lubbers.

The day after Sam's visit, I started packing.

Now I'd managed to cross the country and get the very job Sam had recommended for me. Seymour was waiting for me in our cabin, and I had a fresh wave of fantasy to live up to. It would all work out. It had to. Eyes front.

2

I was raped that same night, at home on my kitchen floor, and raped not with a penis but a steak knife.

While I was in Morgan City, Seymour fell off the wagon into irrational rage, rage that had been simmering under his polite façade. I had no clue it was coming. When I walked in the door grinning, waving my job papers proudly over my head, he exploded. "Don't tell me that when you're surrounded by *men* all night and day you won't be going to bed with them. I know how much *you* like *sex.*"

At first I just tsked. Seymour, don't you know me at all? I'm an unshakable monogamist. And listen about the drunk with the plastic duck diaper pin, the sad hyena brothers. You call those men?

But there was no talking Seymour down. He raged. We argued. Finally I went off to bed alone. He'd calm down when he sobered up.

It started in the bedroom while I slept. He stuffed a blanket over my face and clubbed me through it, with his fists. My screams, my screams.

Next he dragged me into the kitchen by my hair. I saw him rummaging in the knife drawer with his free hand; I struggled harder. He jabbed my face with his elbow. Then I was on the floor, and Seymour had a knife in each hand.

With one of them, a fine Sabatier steel straightedge, at my throat, he backed me to an arched kneeling posture, the top of my head braced against our homely kitchen stove, his spittle flying furious in my face.

"Open up. Open *up!*" He nudged my bare thigh with the smaller knife, the serrated steak knife. "Open up, I'm telling you, open *up!*"

You cannot know about rape. You just can't know until you bleed with knowledge. How your trust in the universe, your belief in your own will, shrinks to a germ of frozen fear, and then goes out like a light. You would not believe, until a knife punctures the skin of your thighs, that yes you would separate them, even to a knife; that you would beg and blubber and piss and shit, that you would search his familiar face gone amok and unseeing to please Please please please don't please Don't. How could you know?

I summoned at one point some grotesque imitation of sanity, of grasp of the situation: "Now, Seymour, wait. Wait, just wait, wait. Let's talk . . ." The prick of his knife brought a wave of bile up my throat, and I am ashamed to say I gurgled through it, "I love you! Remember? You love me! You can't do this!" I did all of that and more, hearing meanwhile careless cars skimming by on the blacktop highway. It was Halloween Night.

I stopped struggling when the little knife impressed its way to my puckered innerness, especially then: no whimper, no scream, just the cold with no breathing. Expecting the worst, and the worst was coming.

"That's what you want, isn't it? A cock in you, that's all you want, isn't it?" Here armlock tightened on shoulders, here knife blade sliced neckflesh so-barely, with such a fine blade. I squirmed. He made the squirming hurt, thrusting up with the little knife, the knife inside.

The rape lasted just a little while. I don't know how long. Just the barest little slice of my life, maybe not even as long as it takes to smoke a cigarette, or pick out a magazine at a newsstand, an inconsequential amount of time.

He did finally stop; he must have seen himself. Suddenly he stumbled to a corner where he dropped in a heap, hugging his knees, knocking his head against the floor. Crying, "What have I done? How could I do it? OhhhhGodhelpme!" In that moment, I could feel his heart rend, hear his universe shock-stop in silent ruins. For a time there, I knew his feelings better than my own.

It's a funny thing, the way compassion took me over then. I crawled across the floor and held him on my knees, stepping out of my own pain, to cry with him. Not him with me, me with him. Odd, isn't it?

I can still feel his tears, wetting the arms I held him in. Every wail from his throat tore at me too. Never in our two years together had I seen him psychically naked. Always was the wily boy protecting the wily boy protecting his secrets one and all. (Relatives of alcoholics will know exactly what I mean by this.) At last I'd seen him: the horror, the pain, the ruin. He had seen this too.

When his tears subsided, my own feelings rushed in, jerking me in spasms to my feet. I'd been holding him; I think I dropped him then. Because I heard a terrible noise that I will never forget: a choking with screams sucked in, a mindless death-rattling sob, and that ugly, ugly sound came from me. I was crying, as he had, the shrieks that pierce the throat, rack the ribs, end all the beginnings, align every moment of life that's gone before into a hideous deathmarch. So this is where I'd been heading all my messy life: to this Now. Halloween Night.

I don't remember that he comforted me. He may have tried. I may have shaken him off. I do know that I was howling down wetly into the inch of space between the cold white stove and the kitchen wall when the *screek* of the screen door opening snapped me up. I turned to see Seymour backing out the door. To take a walk in the pinewoods behind our house, he said when he saw me look up. His eyes darted from one corner of the ceiling to another, avoiding mine. Then he was gone, and my crying stopped. For a while there was only the *cherrr* of tree frogs, the slow, ponderous *drap* of the leaking sink faucet.

I remember next finding myself on hands and knees wiping up the evidence of my shame, my defeat. Yellow shame, brown defeat, flecked with the red of my own fresh blood. Hide it! Quick! That's odd, too, isn't it? An animal reflex, maybe? I don't know, I really don't.

But somewhere in there I must have realized that certain accidents can't be wiped away with Lysol and wet mops.

Next I was cleaning myself and packing my traveling bags. My hands jerked at the ends of my arms; my face hurt, stretched into some no doubt ghastly expression. My thoughts raced and faltered, most of them nonsense word progressions with only my shock to string them together: doctor . . . now wait . . . my wait, wait my home . . . wait is where is my my my wait . . .

But something was left of me, something cool enough to creep outside to my truck, carefully stow a berserk assortment of my personal gear in its empty bed, slip the key into the ignition, ready. Minutes later, Seymour returned, carrying a magnolia branch.

"I couldn't find an olive tree," he said, offering a shyboy smile. It was to be a peace, then? We were to dismiss what had gone before as a bad dream, a non-event? So far I hadn't quite grasped what had happened, but my anger was beginning to gather itself, just threads at first, weak gray threads, not enough to put my weight on, but something.

"That's nice, honey," I said, setting the magnolia branch upright in a pretty little vase in the kitchen window of our pretty little dream-house-cabin-in-the-woods. Meanwhile I searched my memory. I had never done much thinking about rape. Was it still a capital crime? I hoped so. While Seymour soaked in a hot tub, I eased my truck out onto the highway and drove to the sheriff's office.

"How come you callin' it rape?" the deputy asked. "You was livin' with the man, wasn't you?" I didn't answer him, couldn't. He went on. He wanted all the information. "Let me see the marks on your laigs," he said. I still couldn't answer, or move. I wondered, is this the beginning of catatonia?

Finally the sheriff gave up his questioning and drove me to the local J.P., who pecked out a warrant on her ancient Underwood. A warrant for Seymour's arrest on charges of aggravated assault. The deputy said he'd drive on by the house and pick him up, if I really meant to go through with this, but not until the district attorney okayed the warrant. Tuesday, maybe. More like Thursday, the same day I was due on the boats. All through the questioning and arranging, I hadn't cried but only shivered, blue and stuck. But when the information that I no longer had a home settled into me, I felt my throat burn again, felt my eyes fill up. No home, no home.

A female deputy arranged for me to spend the week at a local women's shelter. I unpacked my bag there in the chill predawn. No other women shared the house that night, so I had the pick of beds in which to hug my legs and scream. My thighs as dead as stone slabs, my throat choking voice cracking sore with the rage at the bottom of my fear, alone, screaming into a pillow: It's not fair. My innocence had not protected me, my weakness had not protected me. I could not, it seemed, protect myself.

Later, squatting ankle deep in a clawfoot tub, tasting sulphur in the steam of scalding water, letting my head hang down, feeling more than seeing the bluelight seeping through plastic window curtains, letting my jaw drop loose while the shock took me, I saw: he could have killed me, as easily as that. I could as easily be dead as alive. He is stronger than me and I am lost.

My thoughts spun, desperate to make sense of this senseless thing, but the revelations came too fast; I couldn't absorb them all, let alone order them into a new reality.

Swabbing my cuts with Mercurochrome, slowly slowly peeling paper strips from Band-Aids, I asked myself, in all seriousness, had it been something I said? Could Seymour have been inspired to destroy me by some act of my own, some trigger I'd pulled perhaps knowingly, perhaps not?

It was a long time before I understood that no one deserves to be raped, that life doesn't operate on a scheme of who deserves and who doesn't. But for as long as I believed I'd brought the horror on myself, I was still, in some very small way, powerful. Admitting that I'd been raped by his drunken, deranged will, against my own, I lost it all. I put off that knowing for many months. Maybe I don't even know it now.

Feeling dim and unattached, I swaddled myself in three layers of clothing and stepped out into a narrow little street, Mary Street, to buy a cola at the neighborhood market. It was nearly seven a.m., blue-collar rush hour. I heard the highway traffic two blocks away, wondered how people could just get up in the morning and go on with it as if nothing had happened, as if they had never learned how dangerous was the world. How some people were stronger than others, that none of us is strong enough. I was just barely bumbling along Mary Street, toward the sound of the traffic, when two sleek alley dogs leaped in front of me, black, taut-muscled, barking. When they saw they'd scared me, their rush resolved into a slinking pursuit; they raised their ears and prowled toward me in a half crouch, showing their teeth, ready to spring.

I don't know where it came from, the strength I found to refuse to be hurt again. I stomped my feet and growled, "Back!" My heart was in it, too, not knocking helplessly against my ribs but humming with fury. "Back!" I threw my belly forward, showed my own teeth, blasted two thousand volts of black menace into their faces. "Back!"

The dogs divided in confusion, barking hollowly with the empty bravado of the vanquished. I tossed my head and walked on. Nobody was going to chase me off Mary Street.

3

Five mornings later, still homeless, still nursing myself through shock, I drove back to the boat company's Morgan City depot. I arrived an hour before the 0400 I'd promised. I didn't want to miss my boat.

The carryall van I'd be traveling in was gassed up and ready to go, but what with one mysterious delay after another, it was dawn before we loaded the van to head for our port.

I learned from the supply dispatcher that the name of my boat was the *Harbor Pride,* and that her port was Intracoastal City, Louisiana, two and a half hours away. What we were doing was called crew change. Our cargo for the trip could not have been less exotic: half a dozen string mops and four hundred pounds of groceries.

The *Harbor Pride*'s engineer, introduced as Cupp, loaded his gear into the van: a pair of heavy toolboxes, a quart bottle of vodka, and a change of clothes tied up in a pair of work pants. My own gear was more rightly a caravan: bags and more bags of clothes, a carton of books, my cameras and film, a food processor, and a nonportable typewriter. "Don't b'lieve I ever seen them things on a boat," the dispatcher said, eyeing the last two items. "Yew a Yankee?"

Cupp, a pinch-faced Appalachian type, just winced at my

flurry of questions. So we drove across the southern edge of Louisiana in silence with the sun rising at our backs.

Our road, a two-lane blacktop flanked by overgrown bayous, ran through trailer park towns with drawbridges over the Intracoastal Waterway, through clapboard towns with bastardized French names and neon signs advertising Hot Boudin, past green swampy plains dotted with slow-moving hump-necked cattle. I watched small companies of crows glide out of the roadside brush to raid the carcasses of armadillos that lay split and seeping on the blacktop. I drifted off to sleep reassuring myself that I was bound for what must be a more romantic spot, a genuine port. I pictured bearded sailors with rings in their ears, cargo crates stenciled with the names of exotic cities, open-air fruit markets.

The van stopped with a jolt and I woke up to see cowboys driving scraggly longhorn cattle across the highway. Cowboys? It was as if I'd stumbled into the odd mix of heroes and landscapes of the MGM back lot. But yes these were cowboys, the tanned and dusty real thing, as small as jockeys though, mounted on short-legged ponies whose heads hung low. Louisiana, I remembered then, abuts East Texas. Cupp and I were caught in a roundup. He yawned. I goggled. Another country. Then the cowboys whipped the last stragglers across the road and we drove on. I drifted back to sleep.

I woke up again when the van's wheels rattled over white shell gravel. This was our destination, a square block of parking lot with a twelve-foot Cyclone fence surrounding it on three sides. On the fourth side, water: a wide brown oil-slicked river that I later learned was called Freshwater Bayou. The only structure on the lot was a dented trailer with a long cement block room built onto it—outland oil company architecture. Cupp grumbled something incomprehensible, climbed into the back seat, and went to sleep before I could ask him where was our boat.

Road gritty and impatient, I watched a crew boat backed up to the dock next door and held there with its diesels snarling while forty rigrats, all tricked up in their Sunday urban cowboy best, tumbled off the boat and piled into a waiting charter bus. Then a convoy of eighteen-wheeled flatbed trucks loaded with drill pipe pulled up behind us on the lot. The bow-legged lead driver strutted over to the company trailer with a clipboard load of paperwork. The other drivers climbed down from their rigs, stretching their arms and scratching their crotches absentmindedly. One driver turned away from the circle of his companions and emptied his bladder in my direction. He hadn't seen me. Still, I guessed that either the local piss etiquette differed from the north's, or else women were a rare sight on the docks. Both speculations turned out to be true. I was not quite the only woman in the offshore oilfields. I would soon meet Linda, a ballet dancer turned semi-trailer-truck driver; Honey, the mud dock secretary; Angel, Pearl, and Teapot, all boat cooks. I'd meet Bonnie and Ginger, galley hands on an oil rig; Louise, a crew boat deckhand; Sue, a captain's wife; Kathy, a captains' groupie; Corrine, the one and only female captain. Dot owned a waterfront whorehouse/bar. There were maybe one hundred of us in the oilfields. Maybe even three hundred. No more than that.

Men numbered in the tens of thousands. Derrick riggers, truckers, welders, roustabouts, chopper pilots, crane operators, mechanics, dispatchers, sea captains. Men working days and often weeks out of reach of the opposite sex.

A woman, then, wouldn't a woman be the prize of prizes, the queen of bees? Wouldn't even the least appealing female find love at this end of the rainbow? I hoped so. For all the years of my adult life, I'd only briefly lived without a man of my own. I could not imagine life without a man. I saw the rape as an accident, a lightning bolt that would never strike again. I eyed the truck drivers, the boat crews, the rigrats even, wondering if one of them might be my next true love. And I did meet my true love that day, just minutes later. But like everything else in this foreign country, she was not what I'd expected at all.

There she came, bigger than a barn, churning up the bayou as she backed and rounded to dock just twenty feet from my face. The *Harbor Pride*, my boat. God, she was magnificent. I'd expected something the size of the crew boats, but no, this was a giant seedy queen, rolling on rusty haunches, roaring with power, trailing a deep white wake as wide as a highway. The river's current fought her, hard. She persevered, impervious, a stolid First National Bank of a boat. A ship, actually.

She was three stories high at the "house" end, the bow. Behind the house, a low back deck big enough for a Grand Old Opry. She was built like a big old tramp steamer with its hull scooped out from beam to stern, all the way to the waterline. Overall, she was from here to the sixty-yard line in length, one-fourth that wide. Her scrappy-looking hull was painted several random shades of navy blue, her house a yellowing white trimmed in safety orange.

I could see the scars of her hard service at sea: a fresh dent the size of a full-grown grizzly bear in her flat-ended stern, long and uneven scrapes on her massive hull, slapdashes of orange primer over steel skin. Her once-white railings were eroded in spots, and rivers of rust as dark as old blood ran from the drawn-up anchor high in her bow to the waterline two stories below. She was beautiful, simply beautiful. And suddenly exactly what I wanted for a home.

I saw that on this *Pride* I would not be above and remote from the sea. Her back deck, cluttered with piles of chain, hose, and empty pallets, was so near the waterline that wheel-wash splashed over her deck planks. Bordering the deck, great steel stanchions stood like squat goal posts, looped with fifty-foot lengths of aging gray docking line. Two grimy trapezoidal smokestacks, half the size of railroad cars, jutted out of the deck midway to her stern. I imagined I'd get wet and dirty, living and working on this wonderful boat. She'd be a better place than an apple orchard to play tomboy. I love playing tomboy.

Three of the *Harbor Pride*'s five crewmen were on deck when she lumbered into dock. Two of them, presumably deckhands, cast the thick docking lines overhead to lasso the dock's bits. Their startlingly accurate marksmanship, executcd with elaborate nonchalance, put me in immediate awe of them both.

The other crewman, presumably the captain, stood two stories over the main deck at an exposed aluminum steering wheel, facing the stern. These were the boat's stern controls, an auxiliary steering system used for docking. I didn't know that then. I didn't know anything at all about boats.

The captain looked to be a standard-variety oilfield vagrant, not at all the brass-buttoned captain-hatted old salt I expected. He wore a nondescript cotton jacket and a baggy pair of Levi's. His hair was tied back in a Grateful Dead ponytail and crowned with a red baseball cap. He wore sunglasses, too, the mirror-reflective kind I call Fly Eyes.

When the roar of the *Pride*'s diesels died, Cupp and I crossed the two-foot gap between dock and boat (I felt cheated; no gangplank!) to greet Captain Fly Eyes and the crew.

For openers, nobody said "Welcome aboard." It was my first day on any real boat. I was an alien to water, to my new blue collar, to the offshore oilfields. I would have liked to shake hands all around. I stuck out my hand. No one took it. Embarrassed, I turned my handshaking gesture into a wide sweep and, muddled, put words to it. "I've never been on a boat before . . ." No one answered. Feet shuffled in silence. The men moved off, back to their jobs.

For the first of many hundreds of times in my year offshore, I asked myself what was I doing wrong? Was it my too-new deck shoes? My accent? A female offering a handshake to a male? What could account for this sledgehammer silence?

But I didn't have the time, then, to deal with such intangibles. This was my first day on a new job, in a new medium and milieu. Enough to sort out port from starboard.

The first words anyone spoke directly to me on that boat came from the captain: "Don't do that." I'd begun unloading groceries from the carryall onto the boat. "There's deckhands for that carrying. We don't want you hurtin' yourself."

And so I investigated the galley. Stepping through a round-

cornered hatchway from the deck into the house of the boat,
I was again surprised. The boat's interior was clean and orderly. Her white walls gleamed, her linoleum floors shone.
And the galley, just inside the door, was not at all the dirty,
crowded cubbyhole I expected. Instead there were walls of
storage closets, walk-in freezers and coolers, twenty paces of
spotless counter space. A kitchen fine enough for Maxim's,
and all I had to do was feed five of us three times a day. This
was my first encounter with the crazy luxuries of the oilfields.
I didn't mind it at all. I just dug in to work.

After a nerve-rackingly silent lunch the deckhands went
out to work on deck, the engineer to his engine room, the
captain to the dock's dispatch office. I wondered again what
was behind the men's silence. They hadn't even looked me in
the eye yet. Had the fried chicken been too greasy? The biscuits some unsouthern shape? Maybe devil's-food cake was
considered bad luck on the boats? All my life I'd heard about
southern hospitality. I must be making some Yankee blunder
that short-circuited it.

Weeks later I heard Cupp tell our dispatcher I'd made the
best meal he'd ever had on a boat on the day I came aboard.
But that first day he wasn't talking. No, wait, that's not quite
true. He did speak to me once at lunch, and when I couldn't
catch his meaning, the captain translated: "He wants to know,
are them your natural teeth?"

I was glad to be alone now, so I could explore the boat.

The galley, with its combination dining room and TV
lounge, shared the first floor with a big double bathroom, a
spare parts supply room, and the two staterooms allotted to
the deckhands. Each man had a room of his own, a big room
built to house six men. Below the galley floor was the engine
room. Its generators hummed just under my feet. But I put
off exploring it to avoid another awkward nonconversation
with Cupp.

Instead, I tiptoed up a wide steel stairway to what is called
the weather deck, where the three main staterooms, one of
them mine, were situated. All three rooms were monastic but

spacious, a kind of stripped-down Waldorf with round brass-bound portholes. One door on the weather deck opened into a tidy bathroom, which I must learn to call a head. Another door, more properly a hatch, led to a railed deck that encircled the house. I stepped outside and back in again, trying out the mysterious-looking door catches, which I must learn to call dogs. There were six of these dogs per door, one near each corner, two midway down, and I had a hard bit of figuring to do to loose them in the proper combination. I wondered why even the doors had to be so foreign. Later I learned that hatches were so constructed as to seal out the sea and seal in the air if the boat should go down.

I ventured up another set of steps to what I saw then as being the real boat, a square white room that had no equivalent on land: the wheelhouse. Mahogany paneling framed huge brassbound windows with a long view of marsh islands and brown bayou. Beneath one corner of windows lay a sleeping bench with scratchy wool blankets neatly folded on top. Beneath the opposite rear window, a chart table with framed certificates of registry, a hardwood map rack bolted to the ceiling above. Inside the rack was a row of carefully rolled navigation charts.

At the business end of the wheelhouse was a mahogany dashboard with a great wooden ship's wheel fixed front and center. The dashboard was especially compelling, with its giant compass that looked like a crystal ball, its two gleaming chrome throttles, a row of gauges and radios. A tall captain's chair sat before the wheel, bolted to the floor. To its right was a mammoth radar set, bolted down too.

I smelled seawater, brass polish, Lemon Pledge, and something more: a whiff of hard decisions, nervous sweat, tight corners. I wanted to sit in the captain's chair, but stopped myself, afraid someone would catch me there. A sailor probably had to earn a right to that seat the hard way.

Stepping outside through another dogged-down hatch, I walked the railed deck that circled the wheelhouse. Near the stern controls, another ladder headed up. I climbed it to the wheelhouse roof. Between the huge brass spotlights and the radar antenna was space enough for five people to stretch out and sunbathe; I wondered if they ever did, and doubted it.

When I looked down, I saw that three other boats our size were tied up to us, side by side by side. I could have walked halfway across the wide bayou on boat decks. Below me, our deckhands scrubbed down the grimy diesel stacks with long-handled brushes and, I thought, a certain lack of enthusiasm. Already I would gladly have traded my job for theirs, my confinement to the windowless galley for their sun and wind and water.

On the next boat over, a knot of men gathered around some noisy task, power tools screaming. Their cook was running laundry through a wringer washer by the galley door. The dock's crane was busy, too, lifting pallet loads of hundred-pound sacks from a flatbed truck to our deck.

Two decks over, a chubby little cook with an apronload of onions and peppers waved up at me, grinning. I wasn't going to pass up my first real welcome. I clanged down the outside steel stairs and climbed over bulwarks to meet him on his own boat, the *Gemsbok*.

His name was Pistache (Cajun for peanut), he said, and he invited me into his galley for a glass of vodka and Kool-Aid. Vodka and Kool-Aid? "It don't tell on your breath," Pistache said, wrapping the bottle in two layers of potato sacking and hiding it back behind the refrigerator.

While Pistache and I compared notes on biscuits and beans, his crew drifted in to meet me. They were sunburned and scruffy, young, southern, wary. The captain came, too, down from his room at the top of the stairs with a thick black Bible in his hand. He told me what my own crew had not, that my Captain Fly Eyes was just a substitute. The *Pride*'s regular skipper, a man called The Goose, would be back in about a week. "You a Yankee?" the *Gemsbok*'s captain asked me. "You get seasick? How come you workin' on the boats?"

Yes, I was a Yankee. No, I didn't get seasick. But I begged off explaining myself; it was time to start dinner. I was unwilling to admit to anyone, even to myself, that the *Harbor Pride* was my one and only home now. I was ready to love every ripple in her wake, every rivet in her bulwarks, determined to stay aboard her no matter what. I'd lost everything else.

4

I remember my first night aboard the *Pride:*
After dinner (supper), I cleaned the kitchen (galley), went
upstairs (abovedecks) to my bed (bunk), set my alarm for an
hour of the morning I'd always considered nighttime (0430
hours), and listened to my thoughts racing in time with the
Pride's noisy generators (light plants).

I woke, if I slept, at odd intervals, to bursts of noise and
numbers from the myriad of radios, then again to the whine
of diesels cranking up for a run. The *Pride* surged forward
on the water and then seemed not to make progress at all but
only rock gently from time to time. Her rocking lulled me
back to sleep. I woke again later to the urgent clank of anchor
chain whipping out of the hold. When, anticlimactically, my
alarm went off, I stepped out into the darkness of the back
deck in my woolly robe. I smelled salt, felt a dense wet breeze
tickle my hair into my face. We were at sea, tied up under
Teledyne Movible #3, an oil rig.

I'd thought rigs would be maybe the size of the *Pride*,
mounted somehow on flat wooden platforms that floated on
the water. I'd expected to be able to step as easily from boat
to rig as I did from boat to boat at the dock. I'd thought the
thing would be a blocky building fifteen yards square at most.
Instead I confronted the skeleton of an awesome monster, an

erector set gone mad and grown fifteen stories tall. Rigs, I saw, resemble the chemical plants back on Earth, all pipes and lights and fogs of their own making.

In that perfect blackness of night, the eye-mind played a trick. The *Pride,* steeply overshadowed by the rig, seemed flat and still, a reference point of solid ground under my feet. It was the rig that appeared to tilt wildly overhead, weaving and flashing its dozen ruby eyes, pitching away, then toppling headlong toward our naked deck. And all the Gulf of Mexico besides, lifting us, lapping at our steel sides. That night was so dark that the water was a rumor, a whispered conversation just out of earshot. Goosebumps time. So this is where I live now, I thought, hugging myself against the nonstop shivers. Fancy that.

I climbed the *Pride*'s outside ladders, hauling myself up the rail with both hands. Clearly now it was not the rig but the boat that swayed, and as I climbed I fought to stay upright against the pull of her erratic side-to-side momentum. I felt compelled to climb, to get taller, tall enough to shrink that rig back to human scale. When I came level with the wheelhouse, I saw a small white face at the window there. It was Cupp. I stepped inside, blurting, "It's so—so *big* out here."

He made some remark in a mournful or complaining tone, but I couldn't catch what he'd said. I had not yet got the trick of hearing, let alone understanding, Cupp's speech. He cranked his sparse words over a cud of chewing tobacco, through standard merchant marine tooth stubs, and out the screwed-tight corner of his mistrustful north Alabama mouth in an incomprehensible whine. I shook my head, apologized for being hard of hearing. He made a megaphone of his hands and shouted into my ear, something about rope in the wheel, rope in both wheels.

Wheels? Boats have wheels?

"Perpellers!" he hollered.

Oh, rope in the propellers! Sounds bad . . .

"Doomed."

Doomed? So soon? I just got here!

I learned much later exactly what had gone wrong that morning. Our captain was a careless, overconfident kid, the two deckhands no more seawise. Between those three stooges, the lines that tied our boat to the rig had slipped overboard during docking and been sucked up around the propeller shafts. The *Pride* was caught fast, and in a dangerous spot.

Cupp told me divers might come in the morning to cut us loose, but he said we'd have a storm by then. Divers don't work in rough water. Without power, the *Pride* would be helpless. We might well end up on the bottom.

Wasn't there something we could do?

The first clear word from Cupp: "Nope."

And where was the captain? "Sleepin'."

Being new to all this, I couldn't give our danger the credit it deserved. Its sources were too arcane, its consequences too remote from the earthly world I knew. So I perked some coffee, baked biscuits, fried up a nice mess of eggs and bacon. Doomed or no, I couldn't help humming over the stove. I was on a boat, having an adventure. The danger itself was proof of that.

The crew didn't show for breakfast. I ate alone. Then, as I cleared the plates away, the boat began to wallow steeply, lingering on the outside of a roll before heaving back upright. Soon she was rolling faster, deeper, with a sharp lurch when the roll changed direction. The galley's cabinet doors popped open and swung freely, creaking like the gates of hell. Salt shakers flew past my head, deadly missiles now. I ducked, and ducked again when a loose coffee cup whizzed by me and smacked the wall. In a moment all was chaos, skillets and pots sliding off the stove, the dishes sloshing in the sink, condiments dancing behind the refrigerator door. A bowl of half-congealed bacon grease flew off the counter and smashed on the floor. When I staggered over to clean it up, I saw waves in the grease, moving waves. I was already uneasy on my feet. Suddenly I felt irresistibly sleepy too.

I slapped cool water on my face, then scrambled to secure

the galley's loose paraphernalia. I wasn't really worried, although I should have been. All I knew was that the *Pride* was rocking. Boats rock, don't they? But when I stepped out onto the deck to dispose of the bacon grease, a hard wind drove me down the sloping deck and pasted me to the bulwarks. A storm at sea for real. And here came the rain, sizzling over the water with a gray dawn behind it. The sea was empty, the rig even more awesome than it had been by night. The magnitude of this little adventure was beginning to impress itself on me.

It was Cupp himself, not the sleeping captain, who resolved the crisis an hour later and saved our lives, I suppose. I was back in my stateroom, kneeling on my bunk to watch the storm through my porthole, when I heard the diesels roar in short repeated bursts. Then I heard the captain come out of his stateroom and call up the stairs to find out what the hell was going on. But by then Cupp had worked the ropes free of the wheels and the deckhands had us tied up right. The crisis was over. Cupp shut down the main engines and went off to bed himself. See? I didn't need to worry. This *Pride* would always keep me safe. No one could have told me otherwise. I was caught in a rapture for the deeps, a romance with all things boat.

The boat. The boat was the Mother, sensuous, serene, enfolding. Her massive engines hummed as she rocked me on the bosom of my bunk, stirring translucent memories of prenatal bliss. Lush, extravagant dreams came easy, and no, I never did get seasick. When I felt myself losing equilibrium, I surrendered to the boat. She could be trusted completely. Lowering my center of gravity to roll easy with Mother Boat, loosening the slightest muscle of my face so that the spray of salt water should not so much sting as freshen me, flattening my bare feet against the steel deck to let the diesels' beat thrum through me, I saw and could know for a certainty, from horizon to horizon, that all waves are one body and yet, for their brief lapping lifetimes, perfectly distinct; that the Wave is the Ocean, both form and formlessness, and will rise again.

It was a high. I was hooked.

Young Bach must have had a moment like this, the first time he heard music. "So this is where the good stuff has been hiding!" The good stuff was mine alone at night. The sullen crewmen huddled around the blue glow of the galley TV, or lay in their bunks. I prowled the decks alone, seeing things. Flying fish migrating at the full of the moon, weaving the seas together with their twisting silver ribbon. Crowds of monarch butterflies, gliding with the last of the southerly winds to their winter home in Mexico, were caught for a moment in the beams of the *Pride*'s spotlights. Once I was on deck alone when a storm squall drove past, skittering dementedly, touching the boat not at all. I danced barefoot, plashing on the wooden deck while waves boomed over my feet. I had so many secret pleasures.

Even now I can't think of a place I'd rather be at sunrise with a cup of dark hot coffee in my hand than on the deck of a solid workboat one hundred miles out in the blue water. The props throw up a rooster tail of sun-gilded wake. Porpoises race at the bow, Portuguese men o' war set their fragile balloon sails for the horizon, and all the world of water is a mirror.

Every morning, every night, unfailingly, the sky show. Sunrise scored like sheet music for Tibetan gongs and chimes, sunsets layered of fire and soft smoke, orange and rose and lavender, with rippling silver edges.

And who wouldn't trade her mortgage in a subdivision for my perfect stateroom: a ten-by-twelve-foot monk's cell with an ever-changing sea view, a solitary hardwood bunk with two drawers under it, a desk with bookshelves to the ceiling. Two pillows, two wool blankets, two boiled white sheets, exactly three sharp pencils. I added my typewriter, bolted it in place. Enough. For maybe the first time in my life, I had enough.

All my life I'd added on, collected, surrounded myself with a profusion of belongings, as if they could provide me with belongingness. Now I belonged to no one, owned very little, had less instead of the more I'd always craved, and I was healed. My heart stopped racing. I didn't rush from home to work to recreation anymore; here they were all of a piece. I relaxed; time relaxed, ordered now by sunrise, midday, sunset. Friendly tugs in my belly reminded me of my duties: breakfast, dinner, supper. Even in the absence of human society—the men went right on ignoring me—I had all the home I needed, for the time being.

5

I don't think I heard the word *Cajun* more than a dozen times in my year on the Cajun coast. The coast-correct, if vulgar, term for the Acadian French immigrants to Louisiana is *coonass*. Accent first syllable if you aren't one, accent second syllable proudly, and soften the *a*, if you are. The *Pride*'s regular skipper, Captain Auguste Godchaux, The Goose, was a full-blooded coonass who'd spent thirty-one of his forty-nine years as a sailor.

I was napping in my room, staying clear of the men who'd ignored me all that week, when he came aboard. A rap on my door woke me, and I heard an unfamiliar voice rasp, "Yo' cap'n comin' on now. Get up and met wid him."

I opened my door to a ludicrous figure. It was The Goose. Short, shaped like a squat-legged head of garlic, he wore a baseball cap emblazoned with the heraldry of Guillot's Seed Rice, and an iridescent kelly green polystretch leisure suit. I discovered later that he owned seven such suits, a fresh one for every day of the week, all of them sleazy and stiff, in a toxic rainbow of K-Mart colors. Colors like the smells of Styrofoam and epoxy. The suits, to pass his Buddha belly, had to short him at the ankles. Guste, thus rigidly costumed, couldn't bend to lace his own shoes, so he adopted the expedient of wearing cowboy boots with their zippers splayed open.

Guste's face was crook nosed and double chinned, lit by wise Gallic eyes that had seen, at one time or another, all the foolishness the world had to offer. The pattern of laugh lines on his face would seem to indicate that he approved of it. He was looking at me now. "Dey tell me I got a crazy-crazy Yankee cook, me. I din' b'lieve dat shit. Now look! Turn y'seff aroun'." He spun me. "Lookin' somepin like a college gel to me. You college gel?"

Before I had time to answer, Guste was scuttling off down the hall. "Come see, *ma fille*. Show you how you make dem logs." Then and there he transferred the ship's logs to my desk and instructed me in their keeping. "Don' be catchin' dis coonass makin' dem damn logs no mo', no. Got me some smaht college gel fo' dat."

Life with The Goose was a series of lessons. My second lesson, only five minutes after the one in log keeping, was a course in the preparation of coonass coffee. It starts with Community Dark Roast, a close relative of carbon, and perks at a boil with three tablespoons allotted per cup. Guste liked it best when it was left on the stove to simmer for twenty minutes or more. He trimmed his espresso-sized cup with two heaping spoons of sugar. Hot carbonized mocha pudding is what it was.

Over coffee, Guste told me he'd been an orphan ("so po' we had to jag off de dog to feed dat ol' cat"), growing up in the backwoods and swamps of Louisiana at the mercy of one or another "cousin." He attended what he called American School for less than two years, and that only for the free hot lunch. His teachers spoke only English, Guste only Coonass French. They'd told him he was dumb, and that label still haunted him. "Wait fo' I teachin' you coonass talk. Den you see I ain' so dumb like I soun', no."

When Guste signed his name, it was with an amorphous squiggle. On the infrequent occasions when he was forced to write in my presence, he turned his back to me so I wouldn't see what anguish it caused him. Still, I could hear the pencil points snap under his efforts.

"It was a sight, *ma fille*, me takin' dat test fo' cap'n. Still to dis day can' spell 'horizonto' or no udder big words like dat."

When he caught my pen racing over a bulky letter home, he marveled. "I see dat, but I don' b'lieve. Yo' doin' dat fo' fun?"

When Guste asked for my history, I gave him the whole thirty-five years up to and tearfully including the rape two weeks before. He listened with his head drawn back respectfully, his eyebrows tilted at opposing angles, cheeks and nose twisted to one side, mouth making a sagacious off-center O. I didn't learn until much later that he hadn't believed much of what I'd said. A truthful account of personal history is the last thing an old sailor expects from his shipmates.

I began that day to catalogue Guste's facial expressions. The Tell Me More Child Look involved an elaborate lowering of eyelids and puckering of lips to convey that I was probably bullshitting him, but welcome to it, as when I described a blizzard in upstate New York.

(Guste said he knew all he needed to know about New York. A terrible place, he thought, with people lined up asshole to elbow. I asked him what part of New York he'd seen. "Dat Guardian airpo't dair, I seen *dat* t'ing.")

Guste's expression of delight swept his face into unexpected wrinkles and chasms, lifting his eyebrows into symmetrical peaks. He hooted and wheezed when I told him I'd been surprised to learn that boats have wheels.

I saw the least of his deadly, infrequent anger. With his eyes laid like lead coins under his sagging lids, Guste's jowls went slack and he made no sound but the hiss of inward seething, as when he discovered I was oven roasting, rather than pot roasting, two ducks for that day's supper. Yankee cuisine was no joke. Once he cooled down, Guste promised to teach me to cook properly, the coonass way.

For an example of proper coonass cooking, I give you here Guste's recipe for Sauce Piquante:

"Makin' de roux, nuff fo' cover de bottom of de big black

pot. Cook him good, brought dat color up slow so him color fo' a new shoe sole, till him smell fo' nuts. Put you in dem li'l onion, lil' bell pepp', all chop nice. Puttin' dem cook in dat roux so dey pass in, den so dey pop out onct. Take him off dat fire. Put yo' han's full dese tamate wid de hot pepp, take all yo' fingers, soosh dem tamate in de bowl, let dem comin' soff on you. Cass dem in de roux now. Den put you in dis can sauce tamate, dis paste. Put you in dis pitimous bit red pepp [one heaping tablespoon], dis li'l bit salt.

"Put you dis pot top de udder pot, so he cook slow-slow. Not to taste him, not to look on him, not nebber stir him up till he finish."

How do you know when it's done if you don't look at it?

"When dat pot he sit hebby an' quiet, no mo' make a bump. Open him, see if jus' li'l wet place, jus' a bellybutton dair, den you got him right."

It was Guste who dissolved the wall that the crew had built to exclude me. He broke the icy silence at the galley table with genuine laughter, easy conversation. For starters, he gave us all nicknames: "Georgia" to the sleepy deckhand from that state who slept in a long green-striped nightshirt, "Bad Apple" to the truculent nineteen-year-old deckhand with a genius for malingering, the respectful "Chief" to Cupp, who was our licensed chief engineer. Guste called me "Gros Chieu" (tr.: Fatass) and instructed me to call him "Gros Vent" (tr.: Greatgut).

I was, suddenly, welcome aboard the *Pride*. And not just one of the gang, but the captain's pal. Cupp, who'd known The Goose for ten years, thawed toward me immediately, making us a society rather than a pair, an odd society, though. Today when I listen to the tapes I made of our wheelhouse conversations, I hear Guste's halting English sprinkled with bastardized French, Cupp's still inscrutable Alabama whine, my own *New York Review of Books* vocabulary, all of us interrupting one another to ask, "What'd you say?" and "Howzatagin?"

and "Cut dat big word in two an' spit him out again. Yo' talkin' fo' a po' coonass now."

I remember my first experience of the company of men. I was five years old. My mother braided my yellow hair into tight pigtails, tied a white pinafore over my best blue Marshall Field dress, then let me go. It was a wonder. My father was taking me with him to the opening day of the Churchill Downs racing season. From the blur of that day's memories: the taste of the green-painted grandstand rails, the confetti of torn pari-mutuel tickets, the smells of whiskey, tobacco, the paddock. It was 1948 and the men were all hats, shoulders, and pants, loose-legged gabardine pants with pockets full of cash. My father's friends: a man named Frank who barked like a seal to make me laugh; a black bartender with wet eyes and pink palms who brought me a plate of Saltine crackers and soft yellow butter; a priest out of uniform for the day who gave me two dollars out of his hat to bet on number four in the last race. They called me Lucky; "You were our lucky charm today." I fell asleep on the long ride home and pretended to be still sleeping when my father carried me from the car to my room. I loved men. Now I was among them again, full time, and the thrill was as powerful as ever.

Guste, Cupp, and I were very nearly inseparable. They encouraged me to learn the traditionally male secrets of the sea. They relished my interest. In my free time I followed them like a puppy. Then when I had a meal to prepare, they joined me in the galley and kept me company while I worked. Cupp helped me translate my Yankee recipes into southern-style meals. Guste told sea stories.

He told a funny one about the cook who slipped and fell overboard one night in the channel and wasn't missed until the next morning when breakfast was late. (They backtracked and picked him up on the bank.) There was the tragic one about the Watercraft executive's wife who joined her husband for the maiden voyage of a boat named after her. Her namesake boat, a magnificent vessel with all the most modern navigational aids, unaccountably rolled over—all the way over—while offloading pipe at a rig. The woman was trapped in her compartment, underwater. "Dat woman drowned with thirty-four thousand dollars' worth of jewl'ry on her han's," Guste said. "Dey said dey could hear her hollerin' down dair for two-three hours befo' she drown."

Guste and Cupp reminisced about the Eisenhower recession years when even the oil business went bad and every hand aboard the few boats still running was a licensed captain. "Dat was a sight, *ma fille.*" Guste laughed. "Cap'ns cleanin' out de toilets, cap'ns cookin' at de stove wid aprons on dem. You talk about some raunchy cooks!"

Guste invited me, and even expected me, to join him in the wheelhouse whenever the *Pride* left her moorings. Without my having asked for it, or even thought of it, he pushed me headlong into a full-time apprenticeship. Since diagrams written on paper were beyond him, he demonstrated the sometimes complex principles of navigation with a pack of cigarettes and a lighter. With only canoe experience behind me, I was hard put to keep up with his instructions. But Guste was that rarest of teachers. Not once did he explain a piece of equipment to me if I could take it into my own hands and discover for myself, through often tedious trial and error, how the thing worked. I'd never changed a flat tire in my life, or taken a math course willingly. But Guste had me calculating wind speed and current direction, boxing the compass and plotting our course after one lesson.

Ten minutes later, he put me on the wheel. That was quite a moment, when I stood up three stories above the world and for the first time directed the movement of a ship as big as a building down a narrow bayou.

I turned the *Pride*'s wheel to the left, just to see what would happen. Nothing happened. Maybe I hadn't turned it far enough? I turned it farther. No response. Guste, watching from one side, shook his head. "Now you fucked up good, *ma fille.*" A chill crept over me. What was happening? What had I done wrong?

Then, slowly, ponderously, the *Pride* reared around on her

stern and headed for the bayou's left bank. Whoah! But wait, no brakes! I spun the wheel in the opposite direction, hard, and waited for the boat to react. Nothing. She kept on straight at a shrimper's dock as if she couldn't hear me screaming, "Wait! Stop! How do I stop this thing!"

Guste stood flat on his feet and refused to rescue me. The dock was dead ahead. Just as the *Pride* was about to ram it, she slowed, seemed to think about the consequences, and then began to come around to starboard again. Whew! Her bow pulled out into midstream, picking up momentum, and then— Oh, God!—headed straight for the right bank. I wrenched the wheel back and waited again for her to straighten out. I don't know why I looked behind us—perhaps for some angel to step forward and save my ass. None did. But I did see our wake then, a shameful white zigzag on the brown water. "You'd better take the wheel back," I told Guste, stepping aside.

He did take the wheel from me, but he put me right back on it as soon as we got to the wide open spaces of the Gulf. I learned there that the lag between steering and response can be as long as thirty seconds, as short as seven, depending on wind and current. Guste seemed surprised that I knew so little about boats. I didn't even know that the rudder is in the stern. I was beginning this apprenticeship at ground zero, with nothing to recommend me for it. I despaired out loud of ever learning it. Guste wouldn't let me off the hook. "Yo'll learn, gel."

Guste was especially tickled to be teaching me what no woman, by common oilfield consent, was ever allowed to learn. He promised that after a couple of weeks of lessons I'd be bringing the *Pride* down the bayou and into the locks by myself.

He made another promise, too. He would, he said, train me up for mate. Although I hadn't yet met a ship's mate, I knew it was quite a job. In my two weeks at sea I'd heard the ongoing debate among sailors over which ship's officer had the most difficult job, and which could most easily endanger the crew. Some said it was the captain, master of the vessel.

The buck stopped with him. Others claimed that the engineer felt the greatest pressure. One slip—setting and then forgetting an intake valve—and the boat could go down. But in Guste's view, the ship's mate took the real heat. A mate must not only keep an eye on the engineer, he must also be prepared to pilot and navigate whenever the captain calls him to duty, and supervise the two deckhands, too, hands who were likely to be as lazy and truculent as Bad Apple. Because the *Pride* was working a relatively light job, she had no mate. The mate's duties fell on Guste. "And dat's de wuss of all," Guste said. It was no wonder he was willing to "train me up." Still, I was excited by the possibility. All I'd have to do was stay aboard the *Pride* with Guste, pile up sea time and experience, take an eventual Coast Guard test, and I'd be a genuine ship's officer.

Guste took special pains to make me understand that this would be no sexual tradeoff. He just liked me, me and my greenhorn enthusiasm. He'd grown weary with the boats after thirty-one years as a sailor. I think I helped him to love them again.

An instance: On a lonely night of traveling down the Freshwater Bayou to the Gulf, Guste allowed me to use the boat's powerful spotlights to pick out alligators along the banks. As I poked the water's edge with a ray of yellow light, I caught the flat red reflections of two eyes, four eyes, six, twenty. Real live alligators! Wild ones! Thrilled with this discovery, I kissed Guste on the top of his balding head. He seemed not to notice.

"*Ma fille,*" he said, "I bin thinkin'. Dis not no suckass job, dis cap'n's job. She a fine ol' boat, her."

They call it a Blue Norther. An especially vicious winter storm that funnels down the narrow channels of the Calcasieu marshlands, drawn to the warm void of the Gulf. "Blue" because an eerie bluebird-blue sky lifts out of the swollen gray seas in a thousand wispy funnels across the horizon. You can get chilled to the bone just watching a Norther come up.

"Spooky time," Captain Guste called it. "Spooky, spooky time. Get yo' coonass wet, don't yo' watch out."

Getting your ass wet is a Cajun euphemism for drowning. Drowning is a specter seldom raised in conversation on workboats. But on a Blue Norther night it could not be far from mind.

Even on such a night, fifty miles offshore, the Gulf buzzed with traffic. The little shrimp boats, having waited until the last possible moment to haul in their nets, raced for shelter in the Freshwater Bayou. We lumbered on, holding as steady a course as could be held, dodging windblown shrimpers and ill-marked well caps, the weather rising against us. Guste left me at the wheel to go check over my work on the logs.

I panicked a little as he left. This was only my fourth lesson on the wheel. What if I lost my course or plowed over a well cap? "Well, den yo' fucked up good, *ma fille.*"

Hours later, landfall. But unfamiliar landmarks. Cap'n! Where in hell are we? Guste labored up the stairs, surveyed our position. "Yo' done loss us good, gel."

Then why was he grinning?

"Ma fille, yo' got to get lost. Yo' got to be run' into ev'ry damn dock on dis water. Got to be tradin' paint wid dis one, creases wid dat udder one. Yo' got to run her agroun', yo' got to just fuck up ev'ry damn day befo' *I* call yo' cap'n."

I'd overshot the sea buoy, and Guste knew exactly where we were, but he stood mum while I franticked around the coastline, while the Norther blew up to full howl. At last I remembered I could take a block number from a pumping platform, match it with the chart, and bring us back on course. Guste lost two hours of sleep standing by me, patiently watching over me, waiting for me to grasp, dimly, what was so obvious to him.

Safe inside the bayou, he took back the wheel and ran our big navy blue bow up on the bank to wait out the storm. I flicked on the spotlights to pass my watch time sightseeing.

Our lights sucked the shadows out of the tall marsh

grasses, pinpointing a few skinny, scabrous, wild-eyed long-horns. Startled in the act of grazing, they first glared back at our too-bright lights, then moved off on their tall knotty legs, out of my view. Behind them the horizon was literally afire, marsh burning off in pink streaks and orange billows, and still the night as blue as a blue eye. Could this be Earth?

Boat work, most of it, is teamwork. Even when a job requires the attention of only one man, there's usually someone else standing by. Guste had me for company in the wheelhouse. I was rarely alone in the galley. The deckhands worked together, always a pair. Only Chief Cupp, a chief with no Indians to call his own, worked by himself, down in the engine room. When he saw how eager I was to know the *Pride*, he offered me a tour of his private domain. I put him off. Maybe I believed I'd find dirty, complex *machines* there, that they'd spoil my love for the *Pride*'s clean simplicity. I did know that an apprentice mate would sooner or later have to master at least the rudimentary mechanics of diesels and pumps. And here I was, the kind of female who squeals for help if her bike chain hops its track. I was afraid to discover the extent of my ignorance.

Cupp asked me again, a few days later, if I was ready yet to see the engine room. "I have some bread in the oven right now," I remember squeaking. Guste caught a whiff of my panic and tried to tease me out of it. "I just haven't had the *time*," I lied. It was exasperating to be caught out.

Later that day, without premeditation, I left water trickling to fill my sink and dashed down alone for a time-limited look at the engine room.

The door to the *Pride*'s underground city opened off the galley. Beyond it, down sixteen steel grid steps, I found a high-ceilinged tunnel, a companionway. White, fifty feet or more long, the walkway was arched over with pipes painted into the ceiling, lined with skeins of cable painted into the walls. It was as if I stood inside an egg, and the egg hummed.

There is something in the very architecture of a companion-way that compels you to hurry, as if every trip were an urgent errand. The thrum of engines sets a heartbeat pace, the planking floor bounds a little underfoot, and it's as if you have no choice but to open yourself and let that heartbeat be your own. Or so it was with me.

I forced myself to pause and check out the landmarks. Ten-inch-diameter steering wheels, what I later learned to call gate valves, interrupted the floor in evenly spaced pairs. Red stencil labels on the eggshell walls said: FIRE STATION #3. BALLAST TANK #1. VENT LINE #1. Those simple labels seemed to me then the coded passwords to my new homeland, hinting at the complexity of what I must learn if I wanted to make good. I sighed. My ambition to make mate was sickeningly preposterous. I would never understand even the companionway, let alone the complex world beyond the engine room door.

The faint murmuring of engines that had buzzed my bare feet in the galley grew to an insistent thrum, then a muffled roar as I approached the dogged-down hatch to the engine room. Opening it, I was buffeted by sound, wind, heat, a rush of feeling that can only be compared to the surge of noise and motion from the grandstands when Derby runners round the last turn to pound down the home stretch.

The power! The scale of it! Machines larger than men!

If the *Pride* had been a Roman galley, four hundred or more slaves would have toiled there, chained to their oars, prompted by a crew of tense coxswains, lashed by overseers. Instead, a pair of great yellow sixteen-cylinder diesels, as big as Chevy sedans, labored in thunder, neither sweating nor suffering. The twin generators hulked behind the main engines, growling out power.

Under the vaulted ivory ceiling, mated pairs of pumps, blowers, and compressors drummed on smoothly in unison or at turns. Giant bullet-shaped air pressure tanks hissed and sighed overhead. It seemed to me, new to industrial interiors, that I had come upon not just the true heart of the boat, but the powerhouse of an entire solar system. I sat down on the gleaming, vibrating deck plates to let the wonderment of it all run over me, through me. My fears of inadequacy dissolved in a rush of love.

If I could love it, I could learn it.

By the time I got back to the kitchen, there was an inch of water on its floor, water I'd left running into the sink, moving water because this was not a kitchen but a galley, not a silly land-bound house but a boat. My boat now.

6

I was out on the wheelhouse deck one morning, squeegeeing the windows, daydreaming about how soon I would make my move from cook to deckhand, when I heard a voice call out, "Hello there, beautiful." I looked up and saw a man, probably a captain, leaning on the wheelhouse rail of the boat tied next to us at dock.

He smiled right into my eyes and said, "I love you."

"I doubt it," I said, smiling back with apparent good nature, taking it as lightly as it must have been meant.

"No, I mean it," he called over to me. "I love you, I really do. Why don't you come over here and work for me?"

The first few True Love attacks had turned my head, I admit it. But the thrill was wearing thin with repetition. I grabbed up my squeegee and ducked back into the wheelhouse, putting a wall between us. "Hey, come on," I heard his muffled shouting. "Tell me your name at least. Just your name!"

I think I'd better explain here, before we go any further, that I am not pretty, let alone beautiful. My face is a long flat pan with a prominent chin and an oversized forehead, some cur cross between peasant German and coal-mining Welsh. My skin is white and hard to tan; my eyes are brown, almost black. I can't think what to say about my nose. My hair, depending on the sun for its color, is either streaky yellow or dull brown and I never curl it. Sometimes I remember to comb it. I stand five and a half feet tall on big feet and wide bones. I'm usually ten or fifteen pounds overweight. Strangers often subtract ten years from my age until they get up close where the marks of middle age are clear enough.

I do like my hands, which are large and useful, with long and literally sensitive fingers. I have a nice voice sometimes, dark and female. But my few good features would never qualify me for prom queen in this age of anorexia. Why, then, did so many True Lovers come courting?

Guste, amazed as I was by all the male attention I attracted, put that question another way: "Where yo' keep dem sex magnets, *ma fille?* Ah don' see dem, me."

My being one of the only women in the oilfields was no doubt my biggest plus. But how account for the fearsome ardor of these suitors? It got so I couldn't even carry out the garbage without hearing a proposal of marriage.

I remember one afternoon, when I was trying my hand at chipping rust off the *Pride*'s port side. It was November, but the Louisiana sun bore down, making an oven of my work space. I was absorbed in my work: the noise, the progress, the finitude of it. Sweat ran from under my goggles, streaking the rust dust on my face. Guste was in the galley, cooking up a gumbo supper that was still hours away.

I saw a stranger approaching, a man uniformed in oilfield jumpsuit and safety-patched hard hat. What little character I could read on his unused face alerted me to faint possibilities of alcoholism and diffuse yearning, but not much else. He is Just a Man. Maybe he's going to tell me to move the boat now?

But no, he says if I was his woman, I wouldn't have to be knocking no rust off no dirty old boat. He'd buy me a nice new trailer home and put in a hookup out in the woods by his mama's house. I could watch color TV all day and wouldn't never have to get dirty like this here. He'd take *care* of me, see.

True Love strikes again. The man's proposal is not a joke or even a cheap proposition, not to him. There is both challenge and pain in his eyes now, whoever he is.

I thank him for the compliment and decline his generous offer, being careful not to leave a scar on his sentimentality. But in the oilfields, a polite refusal is as good as a come-on. The man persists.

I take another tack, explaining that I like knocking rust off this dirty old boat. I'm learning my trade, working my way up to captain. The True Lover is only puzzled, not deterred.

I go back to my work, hoping he'll just fade away. He puts his hand on my arm.

Sure, I could scorch my low-rent Romeo with a flippancy, but I am not protected in the southern oilfields by the covenants of civilization. I am hardly in America. Even a staunch pragmatist can get lost, get hurt, in this social nether world which is, if I am to be believed, some variation on the behavioral sink.

I have to pick up my tools and retreat to the galley, abandoning the work I enjoy. Damn.

Later in my oilfield year, I did develop guidelines for responding to True Lovers. Here is one: When in doubt about how to proceed in a threatening social encounter, use the gesture of maximum obnoxiousness to all present. When these scary, lonely men threw their hearts at my feet, I picked my nose.

That's right, I drew myself up to my full height, moved in just a step beyond decent social distance, explored my nostrils with an unconcerned finger and piped, "Whadjew say?"

It was a social ploy I learned the hard way. And most of the time it worked.

7

*T*he Cajun tongue went out of print when the women and men who spoke it, refugees from Acadia, New France, Canada, fled into the swamps of the American South. I imagine they had more urgent priorities than the preservation of their written language.

A language that is only spoken and never spelled breaks down into sounds; the sounds begin to vary from one community to another. Thus today there are countless Cajun dialects, and the people who speak them have only the vaguest idea of how to put their speech into writing.

"How do you spell that word?" I'd ask The Goose during one of our informal language lessons.

"Dair ain' no spell on dat," he'd tell me.

Still, he managed to teach me a few trenchant phrases. *"Je ne voule pas le merde de ton toro,"* translates to "Don't give me your bullshit," and can be inflected to communicate a dozen shades of feeling, from tenderness to outrage. Guste taught me to say "I am the cook on the boat *Harbor Pride*" in Cajun. I asked him what was the Cajun word for deckhand.

"Deck-*han'*," he replied, straightfaced and scholarly.

When Guste's Cajun pals came to call, and to appraise his Yankee woman cook, I understood that when they called me *"fou,"* they meant I was "crazy as shit," and that when they mentioned my *"titons,"* it was my cue to spit a good-natured *"cochons!"* in their direction.

Although I tried to give the opposite impression, I understood very little of their chatter. My best clues to its meaning came from the sprinkling of distinctly American terms. As in: "Something *de le* something *pour ton* bank financing." Or: "Something *dans ma* truck something something l'overpass."

After one such Cajun breakfast, Guste invited me to join him while he visited some of his "cousins" along the docks. We strolled across the dockyard under a stage-set sky of fat blue velvet clouds that gathered in the east. Under our feet, the crushed white shells that serve for gravel on the Gulf Coast glowed blue with impending dawn.

"Letter for Miss Lucy," Bobo, the dispatcher, sang out to me when we passed his office. We joined him inside. "Letter from yo' mama, look like."

The dispatch office, a long fluorescent-lit green-tiled hall of a room, was lined with rigrats sleeping rumpled and upright in red or blue jumpsuits on uncompromising plastic chairs, their boots propped on seabags, hard hats pulled down over their eyes. On the other side of a glass wall, Bobo and the dock crew lounged at cluttered desks.

Guste spied Bad Apple with his feet up on one of the desks and threw him a dirty look. Just ten minutes before, Guste had sent Apple to clean the wheelhouse interior. Apple lifted his eyebrows casually, disdainfully. If Guste hadn't kicked him off the *Pride* so far, he probably never would.

Guste and I were heading out the door again when a voice called out, "Hey! Girl!" Bad Apple's companion, a day-hire crane operator, was talking to me.

"I hear you been driving that boat, is that the truth?" I saw that the man was very drunk. It was seven o'clock in the morning.

"Dass right, she do," Guste answered for me.

"Well, then, what I want to know," the rogue went on, "is how come you want to make yourself a man?" This was not really a question. I sighed. It was a long way from those fat velvet clouds to everyday life in the oilfields.

Guste let his face go slack with deadpan anger and hustled me out the door.

"*À bientôt,* Bobo," I called back over my shoulder. It was that easy, I believed then, to escape the bad-mouth boys of the oil world.

Guste escorted me first to the helicopter field across the road where forty immaculate choppers, painted to resemble yellow-jacket hornets, were filling up with small cargo, rig crews, and sleepy oilfield specialists. While Guste gathered the morning's gossip and yesterday's newspaper in the mechanics' shed, I wandered among the choppers, nearly unnoticed for once. It was maybe too early in the day for sexual magnetism.

And then the first rotor whipped into life. In a few moments the fuschia edge of the sun appeared on the horizon and one after another of the choppers lifted off, swaying unsteadily just over my head as their pilots reared them around to aim for the Gulf. I crouched, then left the field as the nearest hornet to me whirred its wings. The sky was full of them. I'd never enjoyed air travel, but at that moment I wanted to go up.

Our next stop, the downstream mud dock, was no industrial heaven but a forgotten corner of hell. Inside the dimly lit warehouse, moonmen in safety orange jetsuits stood braced behind flaring blue torches, sealing sheets of plastic into waterproof covers for pallets of dry mud, cement, and chemicals. Forklift trucks zipped down the warehouse aisles, whining with unnecessary speed. Pale and no doubt toxic dust eddied around their wheels.

Guste told me the mud company was the property of Lady Bird Johnson. I wondered if she'd ever seen it.

A bulky white-shirted stranger in dark sunglasses stepped into our path, embraced Guste, and kissed him on both cheeks. The man's name, Guste told me, was Bark. Bark smiled a tight, predatory smile down at me.

"Dis mah li'l Yankee gel cook," Guste said, thrusting me forward for inspection.

"Really, I'm no girl," I blundered. "I'm thirty-five years old."

"Young enough," Bark leered, tilting his sunglasses at my *titons.*

The two broke into a rapid stream of coonass talk and I stepped out of the conversation to take a second look at the ghostly doings of the warehouse. I saw a wall of fifty-five-gallon drums marked: ACID! POISON! The drums were stacked on pallets, four and five tiers high, tottering, it seemed to me, on the verge of disaster. The workmen turned as one to grin in my direction, raising their welders' masks to sear me with knowing looks. (You're a *woman.* We know what a *woman's* for.) I backed up and followed Guste into Bark's tiny office.

I was surprised to see a woman there, my first sighting of a female in three weeks. Honey was her name, and she labored over an IBM Selectric in the fluorescent gloom. She wore a periwinkle blue shirtwaist dress and what looked to be a wig imported from Taiwan. During the introductions, she spared us only her dimpled cheek. But when the men left the room she turned to face me. I was the first woman she'd met who worked on the boats, and she wanted to hear what the boats were like. I described the *Pride* at length, and what with my enthusiasm for it, I probably made it sound like the Taj Mahal. Honey was anxious to see it, "but I know Bark won't let me."

"You've never been on a boat?"

"No, never have. But I want to."

"There are half a dozen boats tied up to your dock right now. Why don't you just walk over and visit?"

"Bark won't let me. Company policy is N.W.O.B. No women on boats."

I wondered if Lady Bird knew about N.W.O.B.

I promised Honey I'd cut through that company policy, and

I was sure I could do it. After all, what's right is right, and this is America.

When the men returned, I invited Bark and Honey to join us on the *Pride* for lunch. Bark said he'd come, but Honey couldn't. I pressed her case, but Bark gave me a flat no. Even Guste's teasing didn't budge him.

That afternoon, as my crew crowded around the galley table for baked ham, yams, and a generous helping of Guste's sea stories, I thought of Honey at her desk, eating her lunch from a paper bag. I wonder now if she still allows herself to be confined to that office, a prisoner of the mud dock.

Nighttimes at dock or at anchor, our crew put the hours away playing cards. After initiating me into the mysteries of Boo-Ray, a regional simpleton's acey-deucy, and pointing out the extra wild cards that make the card game spades into Mississippi spades, Cupp coached me in the use of hand signals we would use as partners in card crime. "Ain't the real Miss'sippi spades less'n we cheat," he said. I was up for it. We were only playing for points, after all.

But Cupp's favorite winning technique at cards was psychological warfare. "Gotta rattle 'em," he said. "Get 'em so mad their teeth hurt."

Primed for certain victory, Cupp and I partnered against Guste and Bad Apple. Cupp, loquacious for once thanks to a bottle of "mineral water" we passed under the table, kept the needle in Apple while we racked up win after win.

"This little girl here, I could make a engineer out of her faster'n inny these so-called oilers they send out from the office. She's got common sinse, is what she's got. Like what this Apple done? Pumpin' drill water to that jack-up rig that done axed for drinkin' water? She wouldn't never of done it. Ain't no stoopit *kid.*"

Bad Apple lowered his head, paying meticulous attention to his bad hand of cards. I knew it was bad; Cupp had dealt it.

Still, Cupp was going too far, using me to get the boy down.

"You shoulda seen her down in the engine room tidday," Cupp went on. "Changin' out the oil in the mains. Wuz me knocked her arm and made her drop the lug wrench down the bilges. But she pop right down after it, 'fore I could hitch up my pants. Come up lookin' like that Tarbaby, grinnin' like a cat, oil from here to here. Wouldn't catch Apple goin' down in them bilges to save his own life."

Guste beamed at me. "Don' have to be tellin' her twice, no. I teach her dat deck, she be tearin' dem deckhan's a new asshole, her."

I shot a nervous look at Bad Apple. His sallow face had flushed to an evil shade of purple. I thought it was time to deescalate the psywar, and opened my mouth to try. Cupp stopped me.

"I tole her if Apple could cook, we'd put the apron on him, put her out on the deck."

Guste, changing the subject to cards, swore that "if I was playin' fo' shit, I couldn' get a bad smell, me."

"Vinegar. Jes' vinegar," Cupp swore back.

The radio crackled then with a message from the rig we were standing under. They were full up with diesel fuel; we could be on our way. Our game broke up. It would be a nightlong trip back to dock.

Out on deck, Cupp and Apple rolled up the fifty-foot-long fuel hoses while Guste and I leaned over the side staring into the water in time-honored sailor fashion. The night was still, gentle.

I told Guste I'd been thinking about quitting cooking to work on deck for real. Deckhands had more interesting work to do, more loose time to learn things than the cook did. Here I'd landed myself the least exciting job on the boat, an indoor job. No sooner would I start a piloting lesson than it was time to make supper. My beloved volunteer night watches ended not in soft sleep but in tense preparations for breakfast. Guste himself had told me that decking was probably too heavy a job for a woman. Still, I wanted to try.

Guste shrugged and smiled and said, "So try!"

I may have stopped breathing for a moment. My bluff was called: so try!

Hoping to look as though I knew what I was doing, I strode to the dark of the back deck and began rolling up a fuel hose. I fumbled it once, twice, three times, self-consciously. By the time I had the unwieldy thing coiled, the three men were watching me, standing by in various stages of smirk.

"Now what?" I asked.

"Carry him to de hose pipe rack," Guste directed.

The coiled hose, an ugly black bulk, weighed seventy pounds or more. Cupp stepped forward to lift it for me.

"She want to do, let her do," Guste commanded.

I picked up the hose and staggered the twenty feet to the rack. I had not known, until then, that my own strength would be sufficient. I stood gasping, amazed and proud, at the sight of *my* hose nestled in *my* rack. "Now what?"

"Dass all now," Guste said.

"But isn't this when you pick up the anchor chain?" Stacking the chain as it came up was another deck job, I knew.

"Tell you one thing," Cupp put in with a gust of inexplicable anger. "No woman can stack that anchor chain. Too heavy. Too dangerous."

That sounded to me like a dare. I'm the original sucker for dares. "Oh yeah? Show me to it."

While I followed Cupp to the bow, Guste clamored after me, huffing, horrified. I wouldn't back off. My pride was on the line.

Cupp prised off the manhole over the chain locker while Bad Apple uncoiled the fire hose he'd be using to wash down the chain as it came up off the sea floor. Guste, overlooking us from the wheelhouse, activated the noisy anchor winch. I climbed down into hell.

I spent an hour in hell that night to prove a point, to prove my love to that boat, too, one will against the world (no muscles yet) while a football field length of eight-inch-link chain clanked relentlessly down *clank* down *clank* down to where I labored to stack it, sixty pounds at a time, in the depths of what sailors call The Hole, the chain locker.

At first it wasn't so bad. I could stand firm on the locker floor and swing the loops of chain to each side of me. But soon, like the standard walls-closing-in nightmare, I was having to stack the chain into the spot where I stood, then find new footholds on the slippery chain, hunching my shoulders and bending my back, using not the leverage of my position but the meager strength of my soft arms to sock that chain away.

I didn't know it, but Cupp was standing wary over the manhole, poised to snatch me out by the hair of my head if the chain brake should give under the strain and start whipping off the walls of The Hole at hurricane speed. I didn't hear until later that night how Cupp had once seen a deckhand whipped into bloody shreds by a chain gone amok.

To climb out of The Hole when the last links were stacked, with a triumphant grin on my muddy sweat-streaked face, that's the vision that kept me going while my bare feet slithered on blue gumbo mud and my arms quaked and my straining neck went rod solid and sweat slicked me from head to toe. I'd never really worked, see. I mean, I'd never even changed a flat tire. I've said that before. It deserves emphasis.

Well, I did do it, and I did have my moment of triumph when I set my feet back on deck and grabbed the perhaps symbolic fire hose from Apple to make an impromptu shower for myself. Hot work!

Guste, from the shadows of the wheelhouse, let go a mighty honk of the boat's air horn, stopping my heart with that sudden doomsday noise. As I aimed the fire hose at his windows, I could hear him whooping and wheezing with joy. "Tole you she tear dem a new asshole! Tole you!"

Cupp gave me a nod of admiration.

Behind me, Bad Apple leaned against the bulwarks, ferociously silent. I didn't think of him then, of how he must have felt. I didn't care, really. Now, more than two years later, I think of him often. For one thing, he may still be working on the boats. I'm not.

Guste began, the next morning at dock, to "learn me the ropes," the literal ropes. I headed for the three-inch lines looped over our deck bitts. "We staht small," Guste forestalled me, "an' jus' faht aroun'."

From the paint locker on deck he produced two tangled lengths of worked-soft manila half-inch line, handed one to me. "Dis yo' line," he said. (I have it still.)

"Unloose him," Guste told me.

When I was a kid I owned a pair of Mexican marionettes whose strings got tangled on their first day in my toy chest. I spent what seemed an entire summer of rainy days working to "unloose" them. As far as I know, they lie in my mother's attic to this day, permanently entwined. If my first lesson at lassoing depended on my unsnarling that manila line, I might never get to the lesson at all.

Frustrated, I tore at the tangles, forced the knots, and made a matted confusion of my line before I noticed that Guste's line, by far the messier tangle a minute ago, was now free and clear. How had he done that?

Without a word he took my manila jumble into his small square hands. He worked it like a patient diamond cutter, like a mother, like an anthropologist sifting potsherds from sand. It was easy. The center of the snarl lay between his open palms. His fingers didn't tug at it but moved gently, parting the puzzle without haste. His arms moved, from shoulder to fingertip, in short blunt strokes, shaking the tangles open, shaking the free ends loose.

"Dass how yo' do him," Guste sighed. "Let him do fo' hisseff."

I felt the sudden weight of an imaginary hand on my neck, a cold hand, entirely out of place in that warm scene. I looked behind me. The broken shadow of Bad Apple splashed across the wheelhouse ladder. As I caught his evil eye, he glanced away quickly, pretending absorption in the window he was pretending to polish. Guste had seen him too. "His po' mama. His po' mama. Shoulda save herseff all dat trouble, raise herseff a nice li'l pig to eat instead."

We went on with the lesson.

Guste claimed he'd been a cowboy when he was a teenager: mending fences, herding longhorns from one marshland to the next, taking part in the fall roundups and the spring brandings. That was hard to believe. I'd seen only a few cowboys so far, but I couldn't imagine this tiny turnip of a sea captain on a Marlboro poster.

But then Guste made an easy lasso in his line and caught me in its loop, dead fast, when I skipped across the deck to fetch a pack of cigarettes the wind had blown away. I decided I'd never again doubt the word of The Goose. He was *good.*

As he lassoed the tall deck bitts time after time, I stood just beside him, mimicking his discus thrower's posture, the wide arc of his right arm, his (it seemed to me) exaggerated followthrough. He never missed.

I took my own first twenty tries, and muffed them all. Handto-eye coordination had never been my strong suit. The small ragged crowd of men who gathered at our dockside gate were no help. They called out random and contradictory advice, nudging one another in the ribs and hooting as I missed the bitt and missed again and missed again.

Guste, hanging back, just scowled at them.

I would have liked to quit. Even when the line grazed its target, my loops invariably failed to open over the bitt and drop onto it obediently, as Guste's had.

The winter sun was hot, high contrast. I watched my clumsy shadow, felt my nerves stutter. And then I hit my mark, once.

That was heartening. I gritted my teeth in concentration, sent the light lasso flying, hauled it back, sent it flying again. Pretty soon I was hitting square on two out of three attempts. The Goose grinned at me, his arms folded over the barrel of his belly, satisfied.

Never in the creaking hour before my first success did Guste take the line from my hand or guide my arm like some silly tennis teacher. He handled my training just as he'd handled that tangled line. Loosely, allowing the thing to work itself out.

After supper, we went back to the deck for more line practice. This time Guste hauled a three-inch-diameter line from our offside stern bitt and dropped it at my feet. I could only just lift its awkward forty-inch loop. "I don't see how I can throw *this*," I whined.

Guste shrugged. "Try and see how she do."

On my first attempt to hurl the thing, it dropped on the deck just paces from my feet. *Thud.*

I turned back to Guste, prepared to surrender my feverish dreams of becoming a real-life deckhand. "See?" I whined again.

"Yo'll do him," Guste said. Then he excused himself to go ashore for what he called "a bit lapp" (tr.: beer). Once he was out of sight I nearly put the line back up on the bitt. I could always *say* I'd tried. But then I saw two faces at the wheelhouse windows, our two deckhands watching me.

Can a woman do it? They didn't think so. I couldn't have borne their being right about that.

So I tried. And I rested. And I tried again. My muscles shrieked and my ears sang in agony. Once I managed to hook the bitt by its horn. But I just couldn't seem to get the damned line high enough off the deck.

And then a feeling of entrancement settled over me, some kind of magic second wind that took me beyond my trying, trying, trying. The line that I'd cursed just minutes before was suddenly whistling over my head, circling the bitt, catching with a satisfying *snap!*

I tested this crazy magic; was it repeatable? Yes, and easier every time. My back straightened, found a balance point. My shoulders loosened. I could feel, actually feel, the power of my own leverage working for me. Here I was doing the lasso act that was the mark of a real deckhand, when back home I'd been a failure at Frisbee. Magic indeed.

I rested, then tested myself again. I could do it; I really could do it. Finally, with the stuffing gone right out of me, I curled up by the bulwarks, leaning my hot cheek on a great steel stanchion, waiting for Guste to return. I puzzled over the fact that all my life I'd heard a formidable list of what women, because they were presumably weaker and more fragile, can't do. In a matter of hours I'd turned one big can't into a can. And I wasn't even strong. Were all those can'ts only myths? Or was I maybe a monster?

When I woke up, my cheek was still resting on the stanchion and I had dew on my teeth. My right arm squeaked in its socket. It was nearly sunrise; time to make breakfast.

After breakfast Guste asked me to show him what I could do with the lines. I tried, but the magic was gone. My throwing arm screamed when I lifted the loop, faltered when I cast the line. A miserable performance. But I knew I had it in me. Guste told me to rest my arm and then practice some more. I did, still puzzling out the cans and can'ts of it. When I began to succeed again, I fueled my practice sessions with private reveries concerning The Little Woman Who Could. As I pulled the loop to my shoulder, I'd call up my adrenaline with a dare: "A woman can't do it, hunh?" Then I'd take a bead on the bitt and hurl the line with all my power. "A woman *can.*" If the loop missed, I'd haul it back, embarrassed but determined. When the loop hit and caught, I'd applaud myself with a loud, gloating "Hah!"

Taking on the world in the name of all women, magnifying my tiny triumphs into a wave of forward-march-for-woman-kind, I was only rescuing myself from the rape. I'd lost there the control of my own life, and was desperate to reclaim it. It's easy to see that, now. But at the time I swept insight away with a rush of furious generalizations. I was Woman, rising from under the heel of the Oppressor.

It was during one of those early practice sessions that I first met Fat Jules, a diesel mechanic ("and a damn sorry one," Cupp told me) who traveled from boat to boat for the Watercraft Company.

Guste was off on a visit to the dispatch office when Fat Jules clambered aboard. There was no one at hand to protect me from Fat Jules's jibes, and worse, his "help."

"No, you doin' it all wrong. Th'owin' that rope jist like a woman." Jules grinned, dimpling nastily.

I gave up the line to him, hoping to avoid a scene. He looped it between his hammy hands in a convincing show of mastery.

I'd been hitting just about half my casts. Jules missed his first four. And then two more, clenching his teeth. He cursed. "You need you a new line here. This loop ain't comin' open right." He dragged a different line into play, threw it, and missed again. " 'Nuther sonofabitchin' bad line," he swore. "Excuse my language."

Fat Jules's face was by then yellow with frustration, slicked with sweat. It looked to me like a bowl of warm yeast dough.

He threw again, furiously, and caught the horn that time. Two tries later, his loop made the bitt.

"Nice," I chirped. "How about some iced tea now?"

But the frenzy was on him. He was going to show me and show me good. Hurling and hitting, hurling and hitting, he puffed himself into a febrile hysteria.

"There!" he brayed with every hit. "There! You seen how I done that?" I felt uncomfortable, embarrassed, as if I were watching a domestic brawl, a tangled, secret vengeance. Watching a mirror, too, maybe. I'd been putting the rape out of my mind, stuffing it into the darkest corner of Bad Memories, Do Not Disturb, unconsciously rewriting myself into the role of victor in these practice sessions. But this scene with Fat Jules breathed new life into my fear, put my anger in its place. Jules was big, raw, immediate, ugly.

Finally, winded and vindicated, with eight straight hits to his credit, Jules smacked the line to the deck at my feet. *"That's* how a *man* does the job."

From that day on, I did my lasso practicing in the dead of night.

I did make a point of getting out on deck every day, though. I needed to learn the boat; I needed to get into condition. The men needed to get used to seeing me at work.

One day I was on deck scrubbing down a diesel stack and Bobo Cenac, our coonass dispatcher, wandered over to visit. He stood alongside me in silence for a while, just watching. Finally he shook his head and spoke.

"Girl," he said, "I b'lieve you the onliest little woman out here trying to be a man. Do a man's job, too, and do good at it, look like. What make you be like dat?

"Here's you, all speck up wid old black junk, dirty shoes on yo' feet, sweaty stuff on yo' face. Why?

"Come see, plenny girls workin' in banks no mo' pretty dan you, meet dem nice men ev'ry day. You, you could be workin' fo' a bank if you clean yo' seff up, comb yo' hair, put on pretty clothes. I know you ain' one-a dem coonass wild women, and you ain' one-a dem ol' bulldaggers neither. You look to be some new kind of crazy, all I can figure."

8

The way it seemed to me then, nothing could happen to me on boats. Not even drowning. I'd always been a good swimmer. How could a good swimmer drown?

No, I reserved my fear and loathing for the land, felt solid on the water. Think of all the things that can't happen to a sailor at sea: her car can't break down, her dog can't get run over, she can't overdraw her bank account, get swamped in some enervating love affair, or get raped, either. Or so it seemed to me.

So while the other sailors X-ed off calendar days until their scheduled home leave, I put the land out of mind. Sticky, troublesome land. My belongings, and perhaps my rapist, awaited me in a pine forest north of Lake Pontchartrain. If I left the boat, I'd have to face the ruins of a life that was distant, distasteful, unreal now. I preferred to remain aboard, magically exempt from the disorders of everyday life in order to learn radar, fog signals, stern controls, sea buoys, conversational Cajun. I kept fat dog-eared notebooks of homemade diagrams of pumping systems, diesel mechanisms, weather patterns. I collected sea stories and colloquialisms. I wrote letters to friends far away, bragging about my adventures. (Eat my wake, landlubbers.) To pack it all up because my "relief" was due to arrive was no relief, was exile.

At the last moment before crew change, the relief cook didn't show up. In fact, the only one of our crew who did get away for home leave was Georgia, who shed his nickname, shouldered his seabag, and left for two weeks in the Morgan City barrooms. The others groaned with envy. I smiled, smug. Things would go on as they had been, and that was fine with me.

Georgia's replacement, a "black" man barely tan and liberally freckled, was the first black I'd met in the oilfields. He was young and gentle, loose-jointed, soft-voiced. When he smiled, his lush thickets of eyelashes interlocked, giving him a sleepy look. His name was Fred. Guste nicknamed him Fred Fatigué.

On his first evening aboard, Fred sat shy at the galley table, laughing soundlessly in all the right places. Then Guste told us about Noiro. "Dat cook dair, Noiro, he wuz a good ol' nigger. Cook us up a couple pies ev'ry dam night. Hurrikin come he still be cookin'. He sho' was a good ol' nigger."

"Still a nigger," Bad Apple flashed, daring Fred to stand up for trial by combat. Fred just kept his head down, but I could see a faint mist in his dark eyes. Guste and Cupp toyed with their silverware while I inhaled to a count of four and took on the job of setting Apple straight.

"You don't talk like that and eat at my table, honky."

Apple just smirked. "Yew ain't in the north no more. Know why?"

I bit. "No, why?"

"Cuz yew nigger-lovin' librals made so much of a mess up north, nobody can live in it no more. All comin' south now. Like yew."

I made some disgusted sound and cast a look at Guste. He avoided my eyes. Cupp spat into his homemade spittoon, looking away, too. The television yammered on blindly.

Later that night Fred and I walked on the docks. He told me he'd grown up in coastal Florida where, sure, the thing was always there, but nobody got up in your face about it. Now, on the boats, the crews couldn't wait for him to sit down to dinner before they'd start in about niggers driving their Cadillacs down to the welfare office, sucking up the hard-earned tax money of honest white citizens. The rednecks wanted Fred to fight. He wouldn't. Especially on the *Pride,* Fred's favorite boat so far since there was only one nigger-baiter aboard.

We ducked out of the wind to finish our conversation in the upholstered depths of Fred's T-roofed Lincoln. How, I wondered, could a deckhand afford this luxury on forty-three dollars a day? I was afraid to ask. I did urge him to take up the defense of blacks next time Bad Apple opened his ugly mouth. I sure wouldn't do it for him. I had enough on my hands standing up for women.

Fred declined to play black militant, but he did promise me one thing, that he'd stand back and let me try the deck jobs, something Bad Apple was unwilling to do. "Now, don't kill yourself on my account," Fred said. "But as far as I'm concerned, you're welcome to all the nigger work on the boat."

I had my opportunity the next evening.

Boat crews call it nigger work. Hell work was my name for it: cleaning out the cement-storage tanks that lie below manholes in the back deck. A small crowd of dockies and rigrats formed when they heard I was going to try it. No woman in oilfield history had ever cleaned a cement tank. No woman could. The dockies knew they'd see me come up choking, a quitter. If they hadn't been there, all set to gloat, I just might have.

I carried a homemade trouble light down a swinging ladder into the ghostly depths of my first tank. The tank was cylindrical, two and a half times my own height, three of my arm-spans in width. Its acoustics were peculiar, disorienting. The deck could have been a mile away instead of just above me. My own breathing sounds were magnified, spooky, a perfect match for the creepy, cornstarchy quality of the cement dust.

The dust, the dust. Dust hung in the unmoving air of the tank. Dust coated the walls and lay in a drifty pool on the tank floor. Dust sifted down from overhead, rose in clouds when I stirred it with my footsteps, sucked toward me when I inhaled.

Cement dust is hard to wash off a human body unless you consider the top layer of your skin expendable. So, on Guste's advice, I'd wrapped myself in two layers of clothing, two layers of gloves, with one towel pinned around my neck and another one wound tightly over my hair. But it is customary to go barefoot into the tank, to avoid wrecking shoes, which are even less washable than cement-coated feet. My naked feet squeaked in the cement dust, nerves squirming with that spooky chalk-between-the-toes feeling, then screaming when I put my weight on my arches and encountered the tank's steel ribs hidden under the dust.

The job is this: to sweep down the walls and overhead, then collect the dust into a five-gallon bucket that depends from a manila rope through the manhole overhead. Once the bucket is full, give two tugs to the rope and the deckhand who's waiting above hauls up the bucket and pitches its contents over the side of the boat. Then he lowers the empty bucket into the tank again for refills. It is work and wait, work and wait, endure the body-heated claustrophobia and breathe as little of the airborne dust as possible. The fine colloidal dust worked down my neck, into my armpits, making first squeaky then gritty work of every move.

Cement stings eyes, so I was furnished with a pair of safety goggles. But sweat soon clouded the goggles, ran down into my mouth. I threw the goggles off.

Cement dust is laced with obnoxious chemicals, so I was furnished with a flimsy breathing mask. But the mask slid on its useless elastic band, and did a better job of barring oxygen than dust. I had to breathe; I threw off the mask. My mouth, my nose, my throat and bronchae puckered instantly. I was tempted to call up for a glass of water, but drinking it would mean washing down a quarter of an inch of cement.

It takes less than an hour to clean a cement tank. The work is not all that strenuous; it is only hell work. Six buckets, ten buckets, twenty buckets go up full and come back empty.

Guste, waiting above the manhole, told me to take a break. If I'd rest, I'd only wind up spending more time in the tank, and time is painful there. So I kept up my bending, scooping, sweeping.

I liked having Guste up there, waiting for me, counting my buckets of progress. When I filled the last bucket and climbed up the ladder, I chased him across the deck and contaminated his tidy costume with a dusty hug. We laughed, we wrestled, I had won. Women can. But I wonder now about the boys and men who climb down into those tanks without a cheering section, with nothing to prove, no congratulations when they climb out again. For them it's simply hell work. And I admit it: I might not have cleaned out a mud tank at all, let alone twice, if it hadn't been for Guste's praise and the thrill of being the first woman to do it. That's the truth.

Fred Fatigué was helpful, too. True, I was doing some of his work for him, but he took a merciless ragging for it from Bad Apple and the dockie gang. When they called him Mama's Boy, it must have stung him. But he never rose to their bait, or reneged on his promise to let me learn his job. He was my guardian angel as well. When I climbed the mast to replace a broken lamp, he waited just below in case I slipped. He showed me how to chip rust and scrape down bad paint and then stood by to make sure I had it right. He showed me how to secure a runaway fire hose when its two hundred pounds of pressure could have blown me right off the deck. Some of the jobs I took on, like changing out the oil in the main engines and lugging ten-gallon buckets of used lube up the stairs from the engine room, were in Cupp's territory and none of Fred's concern. Fred stood by all the same. Fred was closer to my daughters' age than mine, black to my white, quiet to my outspoken. And our friendship broke every local taboo, but he never betrayed it.

Fred even stepped back to let me loose the lines that tied our boat to an oil rig, a high-pressure he-man job if ever there was one. Cupp got angry at him for that.

High waves were breaking over the bulwarks and geysering up through the gunwales that afternoon, so I might have picked a better time to insist on my lesson. I was right at the limit of what my still feeble muscles could do, uncoiling the rig's mammoth six-inch-diameter lines from our bitts with the boat roaring and rearing under my feet. Once I'd begun the unwrapping, I had one of those rare second thoughts: what in the world was I trying to prove? The waves were high, gray, and random, making the *Pride*'s hull shudder when they smacked it. I saw Fred just behind me, biting his lip, tense and ready to dash in and save me if a wave knocked me down. Then a wave knocked me down. Fred moved fast, grabbing at my shirttail to lift me up before the wave receded and sucked me through the rusted-out gunwale and over the side. But Guste's voice stopped him, booming out over the deck p.a.: "Ay, Fred dair! Leave dat woman be!" I scrambled to my feet on my own and finished loosing the line, then signaled the rig's crane to lift it away.

I was shivering all over when Fred and I climbed the outside stairs to the wheelhouse. We stood dripping seawater onto Guste's wheelhouse floor for the ritual chewing-out. Guste glowered at us, really angry. "See dat woman dair? If she start to slip, let her fall. If she wear herseff out, let her wear like dat. Don' give no help to her 'less she aks for it. Udderwise, Ah'm tellin' you, she don' tear you a new asshole, Ah will. Now get dat anchor chain up."

It is both fun and infuriating to be the center and symbol of a hot controversy. Let me tell you about that.

Guste and I were alone in the wheelhouse, bringing the *Pride* into dock from an all-night run. I volunteered to tie her up singlehanded. Fred and Bad Apple were sleeping, and since the maneuver takes only about five minutes, waking the boys didn't seem worthwhile. Besides, I wanted to show that I could do it.

Guste brought the *Pride* into position alongside another supply boat, the one Cupp called *JimBob*. (Its real name was the *Gemsbok,* but Cupp, who called those big black dogs "lavatory tree-ers" and his wife's operation a "misterhectomary," could never have pronounced such an alien word, and so reduced it to the familiar. None of us called it the *Gemsbok* anymore.) As I lifted the stern line for an easy lasso to the *Jimbob*'s stern bitt, a flashbulb went off in my face. I missed the throw.

I recovered myself and made a good throw, then wrapped the line onto the bitt. "I brought my camera with me this time," the *JimBob*'s engineer told me. "Nobody believes me at home when I tell them there's a woman doing the job of a deckhand in Intracoastal City, but now I got your picture doin' it." His flashcube popped again as I growled at him.

The thrill of celebrity. This particular fan didn't ask me to autograph the Polaroid portraits he took. But later in the year another sailor did. After a while, half the sailors on the coast knew my name, had heard of that crazy woman actin' like a man. I was their freak and their wonder, a shocking contradiction to a cornerstone truth of their lives: that what they were doing was Man's Work. I heard that some of them called me The Morphadite. Dig it.

I brushed all that aside. Sure, there was some element of a battle of the sexes in my striving. But the real contest was between me and me. Taking one little job at a time, I found each of them possible, often just barely possible after long practice. I was taking my lumps, too, cracks on the head from the deck stanchions (they don't call them headache bars for nothing) and nasty scrapes from the rough lines. My legs were leopard-spotted with old and new bruises graded from glowing magenta to jaundice yellow. My fingernails were ragged, one of them mashed flat and blue. I collected so many minor wounds so fast that I couldn't even have told you where most of them came from. I hadn't lost a finger yet, but Apple promised me I would. I'd even taken a couple of sharp whacks in the breast from rolling cargo and from the elbows of deckhands wrapping lines at top speed in close quarters. My back was sometimes so sore that I had to take my first few steps off the galley bench in what Guste called "dat chicken walk."

But, God, I was proud of what I could do.

Which is not to say that I ever felt confident that I could Really Do It. Really Doing It would mean cleaning a mud tank and stacking the anchor chain on the same day I'd scrubbed down the deck, tied and untied those heavy lines a dozen times, gone without sleep for an emergency night run. Even Guste wasn't sure I could Really Do It. Still, he enjoyed watching me try.

He enjoyed the controversy, too. He liked to shock his Cajun cousins, telling them he was training me up to be a captain. I was his friend, but I was also his prize two-headed calf.

When a crane operator catcalled at me with some nonsense about "Come over here and I'll teach you how to throw that line," and I catcalled back that if he came over here I'd dump his ass in the bayou, it was Guste who got the laugh out of it.

Even Cupp, the soul of reticence, enjoyed the *Pride*'s being the center of all this attention. But once the company found out, he said, they'd put the hurt on. Women are bad luck on boats, everybody knew that. Worse luck than whistling in the wheelhouse, worse even than carrying a black suitcase aboard. There was bound to be trouble, Cupp said.

I was no black cat or broken mirror, I knew that much. At worst I was only a spy on this man's world. And I did see things.

One night, a night that was clear and unseasonably cold, I stepped out on deck to tie our boat to the dark wooden wall of the Freshwater Bayou Locks. I pulled my hood up over my head against the wind, an inadvertent disguise.

A ragged home-built shrimper rested alongside us, tied to the opposite locks wall with its booms at vertical. The shrimper's galley door burst open and two young boys, neither more than twelve years old, chased each other out into the chill night, both of them wearing only bright white undershorts over their sailors' suntans. The yellow light from their cabin spilled out too, illuminating their race across the bow, up onto the rim of the locks wall where they wrestled as they ran, teasing one another in Cajun singsong. They tumbled in a heap at the lockmaster's door and scuffled there until he pushed his arm out to pick up the paper they'd brought him, the locks report. The boys got to their feet and moved off down the shadowy wall again.

But before they leapt down to their boat, they stopped shoulder to shoulder to make water, adding their own to the everywhereness of bayou, channel, gulf, ocean. Their father, leaning on his boat's galley door, called to them and pointed to the blackness of sky hanging over us. They looked up obediently and I did too. Ah, a ring around the moon. And not the foggy halo of an autumn evening but a slim, lace-edged crystal circle inscribed over a third of the night sky with the crescent moon at its center.

This vision did not come and go like a rainbow, but hung there, stamped on the heavens, reflected in the black water.

I asked Guste what it meant. Red sky at night, sailor's delight . . . red sky at morning, sailor take warning . . . ring around the moon . . .?

"Sailors can't remember dat neither."

I wish I could take you aboard the *Harbor Pride* and put you at her helm. Failing that, let me urge you to go stand on the roof of a three-story brownstone and pretend you have a great ship's wheel under your hands. Try to persuade your senses that the brownstone, a boat now, rocks and sways under you. Don't let yourself tighten up and fight the motion; just sway with the boat. Hear the diesels rumbling behind you. Through that, the insistent whine of the radar set. A salt wind breathes through the wheelhouse hatch and ruffles your hair.

Now, let's move this boat. On the dashboard at your right hand there's a matched set of chrome throttles, one for each of the main engines. Push them both full forward. Look back over your shoulder and see how the props churn up heaving rivers of muddy water, how shock waves of wake slap the little boats tied up at their docks. Your boat, your big boat, surges forward now, making way. The power of it tingles your feet, then your knees, now your spine. That's it. You've got it. And once you've got *that*, can you ever give it up?

Guste often gave me the wheel so he could take a nap on the wheelhouse bench. The nearness of his snores was comforting. I was beginning to get the hang of the wheel, but a trip down that treacherous and narrow Freshwater Bayou was an always ultimate test. This was my chance to prove myself as a wheelman or fail the course. Failure in this case could mean bringing on a genuine maritime disaster.

The bayou was often masked in fog by night. In really dense fog, the world beyond the windshields ceased to exist. Those were the nights when I had to count completely on the radar, and on my feeling for the boat. Once I got the radar properly tuned, I'd see in its round green eye the image of the bayou banks, our *Harbor Pride* a blotch dead center on the pupil. Over Guste's snores, I'd hear the *whisssh-koooshhhh* of the air pressure line when I ticked over the wheel to keep the boat running on an imaginary line equidistant from the meandering bayou banks. Extending my concentration the two hundred feet of the *Pride*'s length, I could feel her bow cutting through the water, feel the water giving way in a row of taut pressures. Her stern wallowed under a full load of cargo; I could feel that too. When the little crew boats rocketed around us, the boat and I rolled in double time.

Passing another huge supply boat heading upstream, I'd nudge the *Pride* close to the bayou's west bank, cut the throttles back to clutch and squeak by the passing boat's hull, not twelve foggy feet away. I'd have to make some quick course adjustments then, fine-tuning the steering to compensate for the suction of the other boat's bow, the thrust of its propwash on our stern. Rolling slow in the wake that followed, I'd feel our own propwash suck us down on the bank, feel our rudders strain to bring us right, feel how we settled onto course again.

It got so I could feel the slightest difference in water depth under our hull when we crossed a pipeline or labored through a shallow stretch of the bayou.

Yes, it's possible to be a big old boat, too, just the way good riders are their mounts, maybe more so. The *Pride*'s steel hull was not inanimate then, but an extension of my own sensitive skin. Standing with my hands on the ship's wheel at the boat's nerve center, I forged my own fine connections to the boat, the water, the shores. Once I heard myself singing, under my breath, "King of the Road."

9

It was crew-change morning for real this time. Guste, Cupp, Apple, and Fred were down in the galley pacing restlessly, ready to go ashore for home leave as soon as our relief crew arrived. I hadn't spent a night on land in six weeks, and I hadn't missed it. Just the thought of leaving the *Pride*, of ever leaving the *Pride*, twisted my heart. I was stripping my bunk, crying with my lips compressed between my teeth so nobody would hear me, when Guste came up to say goodbye. He tried to tease my tears away. "Hey, shitass, yo' the first sailor I ever seen cry to be *leavin'* a damn boat."

That did it; I lost my grip on my lips and blubbered out loud. "I don't want to go."

Guste rocked me in his stubby arms and wiped my wet nose with a pillowcase. "It ain't but fo' seven days, *chère.*"

"But you know how things happen. You've been telling me yourself. The company might decide to send you to Mexico, or me to another boat."

"God dog, don't say it. You shitass, you gonna get me likin' you and get used to you. Then you'll take off gone and I'll be sad fo' a long time."

Guste had told me already how he'd befriended another woman cook just the year before. He'd taught her to hold a course and do the logs. Until now he hadn't said much more about her. Now he told me she'd been a Yankee woman, one like me who liked playing tomboy. He said she wrote what he called "pomes" just for the fun of it. Then one day she'd packed and left, telling him that when she got the itch to travel, she traveled. "I got so close to that girl," Guste said, "dat I put in my mind never to do that no more. When she went? It was like I was loss. Dat gel, she was the biggest help I ever had."

I felt a stab of jealousy. "Was she a good cook?"

"Sho', she could cook. Plenny dem can cook. The cooks de comp'ny sent fo' her place? Dey was comin' on here one after 'nother. I couldn't depend them on nothin'. Dey wouldn't answer the radio, nothin'. Her, though, she was a true person, like how you are."

Now Guste's eyes were wet. "Don' you go off an' leave me, gel. I'd be just as sad, me."

I promised I'd be back. How could I have stayed away? But Guste didn't seem to hear me. He was lost in his own dreads, his own sad memories.

It was just sunrise when I stepped off the *Pride* into the company's carryall van for the ride back to Morgan City. Cupp drove and I sat up front with him, chewing my lip. I'd made up my mind to ask the company for a full-time deck job on the *Pride*. Meanwhile the backseat boys, Bad Apple among them, drank beer and sailed the empty bottles out the window.

"Bombs awayyyy!"

"Beyeeow!"

A lone Canada goose skittered out of the bayou when a bottle ploshed too near.

"Almost got that big ugly sucker!"

"Beeeyoww!"

The carryall's radio wavered between Cajun concertinas and the tiresome whine of country and western victim-of-love songs. I watched the sun come up lush and orange over the flat marshlands where Brahmin and Longhorn cattle raised their heavy necks, swaying in a sun trance.

Back at the Watercraft office, I made my way through the wall of noise that marks a crew-change day: the rough, loving curses of old shipmates meeting, the uneasy shuffle of oilfield vagrants waiting for their new assignments, the ringing of a dozen telephones at once, the click of the heels of white-collared office girls. I stood near the desk of my marine supervisor for almost an hour before I got his attention. Others were waiting, too, to plead their special cases.

The marine supervisor was rat-nosed, a smoothie, a perfect model for a loan company collection man. "I want to be a deckhand," I told him.

He pretended he didn't hear me.

"I want to give up cooking and work on the deck," I said, louder this time. I caught sight of Bad Apple among the dozen or more bored men standing around Loan Company's desk. The men's chatter dropped off. The phones rang on. "I didn't think any woman could do it, but I've been trying the job and I do OK," I said.

Loan Company leaned back in his swivel chair and winked at the men behind me, as if I were a stubborn child, or an uppity nigger. Then he went back to his paperwork, and the room to its noise. "We like to keep our cooks . . . *cooks,*" was all he said.

"Even if I can prove to you today, out on a deck, that I can do the job? Maybe you don't believe I can do the job?"

"No, honey, it's not that at all. We just don't want to see you get hurt."

"Isn't getting hurt or not getting hurt my own responsibility?"

"I have a safety man down the hall would sure give you an argument about that."

"Then maybe I'd better go talk to the safety man."

"Suit yourself."

Safety Man, a marine super-supervisor with a private office, was no friendlier to my cause. He produced a pocket Bible and waved it under my nose for proof of the sanctity of home cooking and small children, the glory of God's plan for women. I must have rolled my eyes a shade theatrically.

"I saw you make that face at me, young lady," he pounced.

"Look," I told him, "my husband and I were divorced more than fifteen years ago. My children are in college now. I've paid my debt to society. Now I want to work on your boats as a deckhand, and you tell me I've got to go back to square one and make babies again. The last I heard, I have some rights. Title Seven of the Civil Rights Act guarantees me this job if I am physically able to do it."

Safety Man leaned over his desk and into my face. "This here is a *private* enterprise, darlin', and it ain't Uncle Sam signing your paycheck. I'm the one with the final sayso about who's fit to be a deckhand and who's not. Your captain tells me you're a good cook. The men all like your cooking. So a cook you are and a cook you will stay. We like to keep our cooks—"

"I know. *Cooks.*"

On my way out of Safety Man's office, I spied the able bodied seaman's handbook on his shelf. Guste had told me to borrow a copy. Instead, I stole it. These jokers might slow me down, but they wouldn't stop me.

I'd been sitting for a while on the concrete levee behind the Watercraft offices, smoking cigarettes, feeling my eyes puff up behind my sunglasses as I choked back tears of rage. Impulsively, unprepared for a real battle, I'd lost. I couldn't even promise myself that someday I'd win. Maybe I wouldn't. Ah, that put the hurt on, as Cupp would have said.

A hand touched my knee and I startled from my slump. A white woman in her fifties, a trim and intelligent face who might have come from my own Yankee world, was asking, "Are you Lucy? From Goose's boat? I've heard about you. My name is Corrine Harmon, Captain Corrine Harmon."

I jumped down from the levee, clumsily. I couldn't seem to keep my balance that day. I thought my trouble had something to do with the heat, or my stress, and I did my best to ignore it. Still, I wobbled when I shook the woman's dry, graceful hand.

"Would you care to join me for a meal at Delish's table on the *Coughlin?* He really is an excellent cook," she said, "though none too clean."

I followed that marvelous being, that incarnation of my own ambitions, aboard the *Coughlin,* where we shared a meal of white beans and rice with the rowdy crew. Corrine introduced me all over the waterfront that day, and while I tagged along I wondered why Guste had never told me about her.

It wasn't until we retired to a bunk room of a ship in dry-dock, where we'd cadged berths for the night, that I had a chance to ask Corrine about her own history. She was a Louisiana aristocrat, I found out, heiress to a prosperous fleet of shrimp boats. Still, she'd spent fifteen years with the Water-craft Company, hoping someday to earn the command of an oceangoing vessel. As we unpacked our gear, she told me that on her first mission as captain of a Watercraft supply boat, her engineer had deliberately sabotaged her.

"He brags about it in all the bars, or I wouldn't accuse him," she said. "And it mustn't go farther than this room, what I'm telling you. I was the captain. I was responsible.

"We were pumping diesel to a rig, and the oil company man called me on the radio to ask if we could pump a full twenty-two thousand gallons. I asked my engineer if we had that much on hand. He said we did, and we did, he wasn't lying about that. We must have had twenty-two thousand and one gallons of fuel aboard, exactly. No sooner had we lifted anchor than we ran out of fuel. You can imagine how I felt. It was my first command, and here I'd done *the* most stupid thing. I had quite a time laying anchor again without power. And since the rig couldn't pump diesel back to us, we had to wait out there in the Gulf until another boat came to our rescue.

"I took full responsibility, of course, but the oil company

the boat was leased to demanded that the company relieve me as captain. Not quite the glory I dreamed of for my first command. That was three years ago . . ."

Corrine's face, nutmeg brown from the sun, stood up on prominent cords from the whiteness of her straight shoulders. As we undressed for bed, I noticed the contrast we two presented. She wore a bra. I didn't. Her armpits were shaved smooth as baby's cheeks. Mine were hairy. She was the proper Victorian matron; I was the feminist Young Turk. I had never seen myself in quite that light.

"I believe in the personal presence of Jesus in my life," Corrine said then, "and I know that these trials are a part of God's plan for me."

My head jerked up. "Corrine! You mean you turn your other cheek to these bastards?"

"My faith is my strength," she answered back, her eyes boring into mine as if in search of a similar presence of Jesus. Ooooo.

I backed off, watched her knot her sheets expertly around the mattress of her bunk and stove in an empty can for an ashtray. She was a real sailor, Jesus or no.

I tried again. "Why do they let you be a captain if they won't even let me be a deckhand?"

"My dear, they don't *let* me be a captain. I paid my own way through the marine academy when they barred me from the company school. I took all the necessary courses, one of them twice. Celestial navigation is a difficult, difficult course.

"And I put up with the sneers of the men in the testing rooms when I went for my captain's license. The tests are extremely tough, you know. But I passed at the top of my class, and the Coast Guard gave me my three-hundred-ton license. No company can interfere with that process."

"But you must have been a deckhand once."

"Surely. On my grandfather's shrimp boats, many years ago."

"You never worked as a deckhand on a supply boat?"

"Oh mercy no! That's *awful* work. I'm sure I could never throw one of those heavy lines. I've never even tried."

"All the same," I told her, "that's the job I want."

"Well, I commend you for your ambition, but do you think you can . . . well . . . take that kind of life? It's very hard work, work for young men or machines, most of it. Not for women."

I told her how I'd stacked the anchor chain, how I'd practiced with the lines, how Guste had put me on the wheel of the *Pride* even in the bayou.

"All that in six weeks? And cooking, too! You must be a human dynamo." She laughed. "You must be needing a little sleep by now, too."

"Sleep! I feel so sick. I can barely stand up now. Look at this." I tried to walk a straight line, failed. "I feel like I'm drunk, but I'm not. The room starts spinning when I lie down."

She wanted to know if I'd ever been seasick.

"No, never."

"Well, then, you're one of *those.*"

"One of what?"

"Sea-lubbers, I call them. They're very rare. They never become seasick until they're back on dry land. Then the earth starts whirling. You're landsick, my dear."

What a pleasant diagnosis: landsick! That proved my love for boats if nothing else did. I lay back on my bunk hugging a pillow to my chest. Maybe now I could enjoy this spinning sensation.

Corrine doused her light. "You'll make it, Lucy. I'm certain of it."

"Would you ask for me to work on your boat, then? As a deckhand?"

"My boat? I don't have a boat."

"But you're a captain, aren't you?"

"And they pay me captain's rates, too," she insisted, proudly. "But I'm working as a mate on the *Philadelphia.*"

I'd seen the *Philadelphia,* a tiny tug. It was the oldest boat in the company fleet.

I lay in the dark there for many hours, spinning, before I slept.

Cupp arranged for me to rent a house next door to his on Sixty Arpents Road, not far from Morgan City. I spent a rainy two days transferring my household from one grim ruin of a home to another. The Rapist, as I had come to refer to the man I'd loved for two years, had moved on.

Confined by a cold rain to that leaky little shack at the edge of the swamp, I found that I could neither sit nor stand. I was edgy with aloneness. I had not lived alone, even for a week, in ten years. My landsickness stuck with me, too. The linoleum floors lurched under my feet, and when I lay down, the room spun in sickening arcs. The sensation intensified when I closed my eyes to sleep.

. . . A blond woman with two sharp knives at her belt offered me one of them. I declined. I must go. The boat would be leaving without me. I turned to get aboard, but the boat was drifting away from dock with its lines trailing in the water. Guste looked back at me expectantly from his post at the stern controls as the distance between boat and dock widened. I made a great dream leap, and I don't know now whether I made it to the safety of the boat (the safety of the boat, the safety of the boat) or fell into the water to be chewed up by the props because that's when I woke up.

I spent few nights on land during that first winter of boats. This dream is the kind that came to me when I did. Missing the boat. A numinous dream, poignant with meaning, but what?

After two bleak days, the rain lifted. I stuffed my seabags again and headed out for an exploratory ride along the coast, that lacy edge of the map of Louisiana where towns named Larose, Galiano, Chauvin, and Cocodrie line the bayous.

In that day of truck touring I saw shrimp boats strung together on wide canals, houses teetering on stilts over December water, herons and cranes stepping solemnly through the flat marshes. Carryall loads of sleeping men whizzed by me, heading for their boats. Carryall loads of drunken men whizzed home.

Then my truck died with a sound like nuts and bolts running through a meat grinder. I was in Leeville, at a bare stretch of abandoned mud docks. At least I was on the water. Sailors, I found, do befriend other sailors in trouble. Ti-Jean Broussard, captain of the boat nearest my truck's breakdown, invited me aboard her. She was the *Wilma Naquin*, an ancient but well-kept tug. Her galley was a snug and homey dollhouse of a kitchen. Ti-Jean called me "pitimous child" and offered me a deep bowl of good crab gumbo. Ti-Jean turned out to be an old friend of The Goose, a Cajun "cousin" even. I was welcome, on that account, to ride back to the *Pride* aboard the *Wilma* when she made her weekly trip west the next day. I slept the night on a spare mattress snugged up against the *Wilma*'s little generator. I didn't even mind abandoning my dead truck. I hoped I'd seen the last of land for a long, long time.

The *Wilma* tooled along the channels the next day, running up through little bayous I'd have thought too shallow for even this tiny tugboat, startling wild ducks and geese from their cover. A deer swam across our bow, wild-eyed, lunging for the protection of the opposite bank. "Go get dat damn gun," Ti-Jean shouted to his alcoholic engineer, J.D.

"Din' clean her," J.D. replied, impassive, popping the top off another can of beer.

I believed, that day, that there would be no limits to the world I'd set out just those few months before to explore, and that all I'd meet with there would be fine, just fine. Drifting with Ti-Jean and J.D. through canals and bayous on a cloud of beery, easy hospitality, I watched the sun's rays slanting into the wheelhouse. They were yellow, alive with golden motes, each mote no doubt a perfect universe. It was a good day.

If *Wilma* had been a larger boat, we might have taken a direct route across Ship Shoals in the Gulf. Instead, we sailed the Intracoastal Waterway that cuts through south Louisiana like a back alley through the seedy side of town. The trip lasted two days and a night, with times out to wait our turn at locks and drawbridges, more time out to fuel up, gossip with the local marina owners, take on more beer. We arrived at the *Harbor Pride*'s dock in midmorning of the second day to find that she'd been called out to a workover rig offshore. Undismayed, Ti-Jean cranked up old *Wilma* once more and chugged down the Freshwater Bayou to the Gulf.

I spotted her from a mile away, my beloved, my boat, rolling steeply between tall whitecaps under a low elevator rig, tied to a dirty yellow pumping platform. She seemed to me for the first time small and defenseless looking on the spread of blue seas. Sunlight sparked off the waves. The *Wilma* bounded underfoot. Spray slapped our faces. "This is the *Wilma Naquin* calling the *Harbor Pride*. Come in, Goose," I spoke into the mike.

Guste's gravel voice shot back. "Dis de Goose talkin'. Who dat purdy woman callin' dair, Ti-Jean? What fo' you steal mah Yankee cook?"

"Gon' give it back now, me. Got me bellyache enuff on dat Yankee poison."

As we neared the *Pride*, I picked out Cupp's pinched white face, his shock of fright-white hair. I could almost read his lips. He'd be saying we were doomed. It was a tricky maneuver we were trying, backing *Wilma* to a rendezvous on the lee side of the *Pride*. *Wilma* slid wildly in the troughs, her scoop-nosed deck awash with slapping water. Turned stern-to the seas, *Wilma* roller-coastered on the wide-backed waves as her single meager engine fought to make way in reverse. I took the opportunity to sail my seabags across to Cupp when our two wheelhouses tilted eerily near, at complementary angles. Then I stood at the rail, awed, gaping, thrilled to the bone at the sight of two boats bucking the ocean within feet of one another. This was a new one on me.

"Ay! Get you back offa dat rail," Guste cried from above me just as a monster wave smashed me against the *Wilma*'s wheelhouse door. A door dog caught me squarely in the solar plexus, but when I got my breath back, I howled with joy. This was the real stuff.

By then the seas had sucked the two boats apart again. J.D. struggled out onto our deck lugging my typewriter, jerking his head to indicate that I should follow him to the stern. He stumbled once with that clumsy load, but I steadied him. Then I slipped and sprawled flat on the deck. The two of us were soaked to the ears, drunk, laughing, riding wild sea monsters that roared and bucked on hard, high waves. Playing with boats.

Soon the rising waves spilled the two boats stern to side. I stood on the *Wilma*'s bumpers, clinging to her rail with one hand, relaying the typewriter from J.D. behind me to Cupp on the other side of the deep-sea chasm. Saltwater geysered up between the two boats and slapped me, hard. The wind whipped my hair into my eyes. And for once the noise of the sea itself drowned the roar of the diesels.

I wish I had a bird's-eye picture: two skinny, stove-in half-drunk sailors straining to transfer an ancient upright type-writer from a tiny tugboat to a massive supply vessel twenty miles out in the Gulf of Mexico under bright sun on a star-tlingly blue sea. The wet woman between the two men, be-tween the two boats, has met her match, her fantasy.

10

*G*uste was as disappointed as I that the men at our company office had refused me as a deckhand. "Whistlepricks," he called them. "Dey won' make no nevermind on you. Ah train you up de same, *ma fille*. You want to learn somet'ing, put you dem pots off de stove and learn her."

We were on our way to another rig, and Guste told me I could go up on it with him that night. This was to be my consolation prize.

To ascend the sixteen stories to the rig, Guste and I climbed onto what is called a personnel basket. The name of the thing is misleading: basket sounds safe, like cradle. This thing in reality is not much more than a kapok inner tube suspended in a fishnet that is suspended in turn from one hundred and fifty feet of one-inch crane cable. To "ride in the basket," you set the toes of your shoes on the inner tube and cling to the outside of the moth-eaten net with your fingers. And if you fear God, you pray.

The crane lifted us off our deck with a jerk. Guste stiffened. I stiffened. Two sailors out of their element. He smiled uneasily as if to say, trust me, everything's okay. But fear clenched his teeth, froze his voice in his throat. This was the first I had seen of Guste's fear. I felt my own, my eyes uncontrollably wide, my shoulders gathered up into my neck waiting for my respiration to click on again while I clung to the swaying net and the net mounted higher into the sky. Swinging above the *Pride*'s rolling deck, a deck that grew tiny in perspective with alarming speed, I clamped my eyes shut and reminded myself not to look down again.

In a minute or two we'd be hauled up to the rig, but for that length of time only the narrow purchase of our toes through the net and the grip of our tense fingers prevented our plunging back to the now postage-stamp-sized deck of the *Pride* sixteen stories below. Oops. I was looking down again. Then Guste shifted his weight and the net swung crazily.

The rig workers, who made this kind of trip often enough, gathered around the crane to ridicule our tight-ass posture on the net. I'd seen rigrats ride down to the boat with one foot and one hand on the net, their opposite extremities hanging loose, making mock of the fear I felt now. "Trust your feet!" one friendly rigrat called out. There was an idea: trust my feet. I told myself I'd look up that rigrat philosopher once we touched down.

But when the crane deposited our basket at the top of the rig, on the helipad, true vertigo replaced frozen fear. The wind up there was hellish. It was as if we were standing on the Empire State Building's observation deck without the customary safety rail around it. Helipads, which look like giant trampolines, don't have handrails. A chopper could get hung up on a handrail.

I could hardly avoid looking down as I crept toward the ladder at the edge of the helipad. Where else was there but down? I saw our *Pride*, bobbing crazily on seas that I knew were not even rough. No wonder rigrats dreaded riding on the boats; ours looked most insecure from this vantage point. Dizziness overtook me again, so I looked straight up, fixing my eyes on the stars—they seemed a world closer—and inched my reluctant toes toward the edge of the helipad. *Oooo-EEE!*

We made it down to the rig and its comforting rails, but the floor there was a metal-mesh deck, thrumming and twaddling as we walked on it. I knew I was too bulky to slip through the half-inch mesh, but my knees wouldn't take my word for it. Only dark nothingness was visible beyond the mesh, down there. Trust your feet, trust your feet.

The oil company man, who'd summoned Guste to the rig to talk over a "little trouble" they were having, led us through a maze of grease-smeared rigging. Up close, the thing looked more than ever like a giant erector set. He showed us how the derrick had been bent from drilling into coral, and I noticed he paled when he told us how close they'd come to cratering out. How close? Maybe ten minutes of drilling, at most an hour. Cratering out is what happens when a cavity on the sea bottom opens up around a platform leg. The rig tips, tilts, smashes over into the Gulf. The oil and gas held back under pressure by the well caps erupt into flame. Death or worse to all forty rigrats and the sailors whose boat is tied to the rig below.

The oil company man explained that the cause of this imminent peril lay in the massive drill rigging. Just two-tenths of an inch out of true at the rig's crow's nest, that almost invisible miscalculation multiplied into a deadly difference four hundred feet below at the wellhead. Guste whistled, or tried to. I asked how soon we'd be going back to the boat, the safety of the boat. But Guste and the oil man had some paperwork to do since it was the *Pride* that would be transporting the dismantled rig to port for repair. While they hunched over their papers, I invaded the rig's "house" for a look at how the other half lives. It was not pretty.

As on the boats, the rig's air conditioning was oppressively chill. But unlike the boats, the rig's house was built of trailers, strings and stacks of dub-l-wides made into impersonal eight-man bunk rooms. Bauhaus in the extreme. The rigrats, when they weren't collecting overtime, collected sack time. That's all there is, on a rig. Their cafeteria was open twenty-four hours a day, a galley hand told me. I asked, "Is it always like this?" By "this" I meant booming with the noise of a pair of color televisions and a big portable tape deck whining out the ever-present country and western crossover hits. There was a rowdy card game going, too. The galley hand puzzled over my question. "Guess so. Nothing special about today but you bein' here." The rigrats ignored me studiously. That suited me fine.

I searched for the mustached face that had called out to me to trust my feet, but I didn't have time to find him. Guste finished his paperwork and we headed back to the helipad for our return-trip thrill ride. He was as anxious as I was to return to the safety of the boat.

I was back on the *Pride,* perfectly content to remain there forever, for one more week before a gallbladder attack put Guste out of action. He tried to bull his way through it, but I saw the gray pain that ringed his eyes. He left the boat on a stretcher in the middle of a night. I offered to go with him to the hospital.

"No no!" he croaked weakly. "My wife wouldn' onner-stan'."

When I got up to make breakfast the next morning, there was a stranger in the galley, Wesley Dumas, interim captain of the *Pride.* Wesley, who preferred to be called Captain, sir, had just recently been licensed to run a supply vessel. That perhaps explains his inordinate fondness for the perquisites of rank. This was my first encounter with dead earnest captaincy, and after Guste's benevolent rule, it was a hard pill to swallow.

Captain, sir, expected his cook to work with an apron on, not to mention shoes on her big feet. He liked his eggs sunnyside up, basted with hot salad oil so their edges crackled like plastic lace. He wanted his potato salad onion-free, with its chunks of green pepper cut no larger than match heads. He liked his T-bone steak chicken fried, and well done. I discovered these particular captainly preferences the first day, by humiliating trial and error. "You callin' this aigs?" "You callin' this potato salad?" "You call *this* a *steak?*"

I wasn't half as deferential and motherly as he would have liked. When he wasn't pouting over that, he was posing for the cover of *Great Captains Annual*, offering his chiseled profile to the sunset. He couldn't have been more than twenty-six years old, but he sure had developed some dignity.

I'm going to try to say something good about him.

Okay, he was handsome, in a gray-eyed, cleft-chinned, homo first sapiens later way. Surely, behind his well-polished authoritarian façade there must have lurked a fully human being. The only clue I got to its existence was his dependence on soap operas. He delayed one semiemergency run to a rig so he could catch the conclusion of an episode of *General Hospital*. Otherwise, his posturing was seamless, easy to hate.

After supper that first evening I spent an hour with Cupp in the engine room helping to install a new water pump. When I returned to the galley, I found Captain Wesley standing over the coffeepot raising an ominous eyebrow at me.

"Just where the hell you been when I need a cup of coffee?"

"There's fresh coffee in the pot." I tested it, puzzled. "It's hot . . ."

"And who's gonna pour it for me?"

"We just generally help ourselves to the coffee."

"Well, we is me now, Miz Yankee Woman, and I'm gonna put it to you plain. I ain't no pussy like that Goose. He been actin' like a woman, from what I heard. Him on the stove and you on the wheel. That shit don't go on no boat of mine. I'm the cap'n. You the cook. Cook pours the coffee for the cap'n. And I don't wanta hear about you workin' down in that engine room no more."

Flashing a black look, I slammed a cup of coffee onto the galley table and slammed myself upstairs for a shower. It's only temporary, I reminded myself.

I was lost in the thunder of the shower for only a minute when I heard a knock on the door, the captain's voice: "Open up."

"I'm in the middle of a shower," I called out. "Use the downstairs head."

"Open up."

I wrapped a towel around me and opened the door a crack. "What do you want?"

"Just wanta look at your pretty little titties," he leered, apparently in good humor again.

I slammed the door. Good grief!

Whenever I got really mad, at Bad Apple's racism or at a line for bruising me, I'd go to the rudder room, the noisiest and most isolated corner of the boat. Behind the harrowing racket of the light plant, I could sob or scream or stamp my feet unheard. Blow it all off. That's where I went after my shower.

But soon the rudder room hatch opened and Captain Wesley poked his head inside. "Been lookin' all over for you. What you doin' in here?"

"Screaming."

That answer made him blink. Wesley may never have heard of Arthur Janov and the Primal Scream. "I thought I told you don't come down here no more. Come on up. I need me some fresh coffee."

Later in the week, when I carried a dinner tray to the captain in the wheelhouse, I overheard him sneering about me to Bad Apple. Something about how "them Yankees bein' so big for they britches" and about "takin' them down a notch." Bad Apple snickered. I was getting my comeuppance.

The captain raised the checkered cloth that covered his dinner. I'd made a special effort that night. "What's this shit?" he asked.

"New England boiled dinner," I answered, lifting my chin.

"Well, this ain't Noo England, and down here we don't boil our damn dinner." He opened the wheelhouse door and hurled the food out into windy space, tray and all. "Now go make me some white beans and rice."

I took the outside stairs back to the galley. The boat was making way on the Gulf, cutting an even blue-white wake through the night seas under a sky as clear as a plate glass window. On such a night the stars are not white and static but reeling in solemn confirmation of the literal truth of the color spectrum, the three dimensions. On such a night the original human astronomers must have lain on their backs in the coarse grass of high altitudes to chart the heavens. I could have been charting the heavens myself. Instead, I was on my way to make white beans and rice. Boy, was I pissed. But, I reminded myself again, this Wesley was only temporary. All I had to do to survive him was act like a real captain's real cook.

It wasn't that easy.

On New Year's Eve, when I'd had a few nips off Cupp's "mineral water" and retired early to avoid seeing 1978 go out, I was awakened by a peculiar but still familiar small noise in my room. A zipper unzipping. Fat Jules of the lasso lesson, drunk now, was waving his silly soft penis at me. I sat up and blinked at him stupidly. He fled. I locked my door and went back to sleep.

The next day Jules, normally an infrequent visitor, showed up for dinner in our galley. He didn't say a word to me or anyone about his bizarre behavior of the night before. Probably he thought I'd let it slide by. Maybe he thought I hadn't recognized him in the half dark of my room. He shouldn't have pressed his luck. After being bullied all week, I was just too tempted to pass on the pain. I ragged him, playing it wide for the crew. "Hey, everybody, here's the famous Pink Flash in person. Come on, Flash, why don't you haul it out where we can all see it."

I never would have imagined that that silly incident would end my career with the Watercraft Company, and banish me from the *Pride* forever.

But Jules had bragged all over the dock about how he'd "been to my room" the night before. And his daddy, Fat Jules Senior, was some kind of board member at Watercraft.

Ragging Jules was not my only actionable offense. When the marine vice president called me into the company office, he told me I'd made a lot of enemies "out there." One of the deckhands (Bad Apple) complained I was trying to take over his job. Captain Dumas himself said I was "uncooperative." Fat Jules wanted me run out of Intracoastal City for holding him up to ridicule. "As the marine vice president, I have no choice but to terminate you."

Fire me? I was shocked. I had never been fired in my life. I tried to tell my side of the story. The marine vice president buzzed his secretary, and she showed me out.

By the time Guste returned to the *Pride*, unsnarled the conflicting stories about me and persuaded the company to reinstate me, I was long gone to another company, another boat. The boat was smaller by far, a crew boat rather than a supply ship. But I was at last a deckhand, a genuine full-time ordinary seaman.

Part Two

WONDER WOMAN

11

Let's say your name is Bobby Lee, Bobby E. Lee Giddens from Blount County, Alabama. It all started with your dad. He come home early one day and seen you passed out in bed stead of bein' in school and he drug you all the way down there. Just so happens you was 'ludin' out at the time and the school nurse got onto it and they kicked you outa school the same day. You coulda gone back but fuck 'em, right? I mean, you don't *need* this shit.

It woulda been all right after that, livin' at home, but then your dad wouldn't allow your mama to set your supper on the table 'less you went and got a job. So you pick up twenty hours a week pumpin' gas at the Do-Rite Station and everything settles down. Next thing you know your boss puts in them new self-serv pumps and you're out of a job again. But it all turns out 'cause your buddy Dwayne quit school just about then and the both of you hitchhike over to Morgan City where they got all the jobs you could ever want.

You and Dwayne was meanin' to work at the shipyards, they make the most money over there and train you up in weldin' and all, but turns out there ain't no place to *live* in Morgan City. Ain't that the shits? Nothin' but forty-dollar-a-night motel rooms. Be spendin' half your paycheck for a fuckin' motel room and the color TV don't even work right half the time. Then Dwayne starts talkin' about workin' on the boats 'cause you get your room and board for free. So you and Dwayne get yourselves some merchant marine I.D., those Z-cards; and first day out lookin' you got yourselves a job, too. But the company sent you and Dwayne off on different boats and the money ain't near as good as you thought it'd be. You don't even know where Dwayne's at anymore. Could be in Texas by now.

And now your captain caught you sleepin' when you was supposed to be watchin' the wheel on a night run. Sonofabitch captain works your ass off all day and specks you to sit the wheel watch all night when all he does all day *and* all night is lie up in his bunk. Shit, they got a automatic pilot on that boat anyhow. Boat ain't gonna get lost or nothin'. Cap'n's scared you'll run into a tanker or somethin'. But them other boats see us comin' all right and go around us right of way or no right of way, you don't care how the hell big they are.

Still and all, the captain runs you off, and if you knew where Dwayne was at you could hitchhike around with him and spend your pay but, shit, might as well get drunk. Find some women, maybe. Turns out, though, there ain't no women around here neither. This here town is the asshole of the universe.

So you're tippin' a few down at The Fiesta and some dude comes in the door hollerin' Anybody here got their Z-card? You say yeah and next thing you know you got another deck job.

At least that's what Bobby E. Lee told me when I asked around the docks for a lead on a job. He told me to go on over to The Fiesta and do what he done. Did.

I got two steps inside The Fiesta and backed out again, the focus of a hundred greedy eyes. This was no place for a woman. So I tried the next best system, the old boys' network. A dispatcher friend of Guste's told me, "Talk to old Curtis down at Tenneco." Old Curtis sent me to old Jerry at Kerr-

McGee, who passed me on to old Jack, who gave me a pork chop out of a bag and let me use the shower in his dispatch trailer to freshen up. Then he sent me on to old Chick. I was sleeping on a lounge chair in Chick's dispatch room when a crew boat captain offered me a ride to Cameron, Louisiana. There was a crew boat over there that needed a deckhand.

"They need somebody real quick," the captain said. "Just don't tell 'em I'm the one sent you. They see I sent a woman and they'll be all over my ass."

It was late that same night when the boat dropped me at a crew boat dock in that foul-smelling smidgen of a dock town, Cameron, Louisiana. Cameron is just across the Calcasieu River from Texas, too close for comfort. I'd promised myself I'd never set foot in Texas. Louisiana was bad enough. Texas could only be worse. But I liked the looks of the boat that needed me, the *Buck*. He was sharp, and new, one of the few pretty boats I'd seen in the oilfields.

I liked the looks of his crew, too. These were no Bobby Lees but certifiable cuties, young guys who looked like transplants from college towns. They were leaning back passing a jay in the *Buck*'s passenger lounge when I stepped in.

"It might be fun to have a girl around," one of them said. "It's heavy work, though." They all looked me over skeptically. I'd put on lipstick for this job interview, and now I regretted it.

But Guste had taught me a parlor trick that proved my strength. (And listen, don't try this one unless you have a healthy back.) I'd bend my knees, encircle a standing anybody with my arms, making sure I had him just below his center of gravity. Then I'd lock my hands left over right and lift him almost a foot off the ground. It's just a loony trick, but an impressive one. I demonstrated on the *Buck*'s one-hundred-and-sixty-pound engineer. The Buckeyes hired me on without a second thought. "We got us a Wonder Woman," they said.

The second thoughts were all mine. Maybe I'd bluffed my way into a job I couldn't handle. I'd had only a spotty two-month apprenticeship at sea, and now no Guste would introduce me, teach me, protect me if worse came to worst. I'd be going it alone.

True, the *Buck* was only a crew boat, not quite one hundred feet long to the *Pride*'s two hundred, one story tall to the *Pride*'s three. Crew boats like the *Buck* have no mud tanks to clean or anchor chain to stack, and their lines are two-inchers, not threes. All a crew boat does is ferry rig crews and light cargo from shore to rig and right back again. They rarely even tie up under the rigs since they're only there for about an hour, compared to the *Pride*'s average stay of two and a half days. I'd even heard rumors that two other women on the coast worked deck on crew boats.

But the hours would be long and the routine unfamiliar. I'd be doing some nasty stevedore work, too. What's more, I was the *Buck*'s only deckhand.

My second thoughts sprouted thirds.

The way I figured, if I didn't succeed as a deckhand, I'd have to go back to a cook's job. No more wheel time, no more plotting courses and watching for sea buoys, no more playing cowgirl with the lines or hosing down the deck watching the sun come up. I'd be exiled to some windowless galley where the fluorescents buzzed and the soap operas blathered all day. White beans and rice, red beans and rice. Horrible thought.

The result of all my apprehension was an ongoing adrenaline trip. I simply frightened myself into hyperperformance. I *had* to make good. Never did I tie the boat to dock without feeling a clammy anxiety, a rush of Gotta Do It. I didn't believe I could afford a mistake, not even one mistake. Not if I were Wonder Woman.

When I washed down the boat, I cleared the between-bulwarks drains with my bare fingers. When I stowed the life preservers, I arranged them symmetrically. I was a compulsive line coiler and wheelhouse tidier. It probably goes without saying that I polished the *Buck*'s brass every day. I had to be perfect. Practice makes perfect. I practiced pull-ups on the ladder handrails, building biceps. I practiced slinging the docking lines in smooth arcs, practiced flying onto the high deck from the low dock in one graceful vault. I learned the trick of hoisting hundred-pound sacks of dry cement from the dock onto my shoulder, then flipping them end over end to land on the deck. The men kept calling me Wonder Woman and I had never felt so isolated in my life.

Stress of the first order does a funny thing to me, maybe to everybody. It turns my conscious self into a cloud, a blur. Through the blur I'd become, I couldn't even feel how isolated I felt. I just kept going, and kept going, perfecting my performance and my isolation.

When anyone offered "help," I refused it. One snide rigrat crew boss snatched the line out of my hands to "help the little lady." I crushed his instep and elbowed him out of my way. Motherfucker.

The rigrat backed off, looking wounded. "I jist tried to *help.*"

Help was one thing, "help" entirely another. By that time I knew the difference. The real thing was infrequent, businesslike. All a real helper ever said was a simple declarative "Here." A "helper," on the other hand, used terms of endearment, adjectives of diminution: "Wait up, babe, let *me* get that." "Move over, darlin', I'll do that little thing for you."

One of my duties was to clean up the passenger lounge when the seasick rigrats disembarked. No one ever offered to "help" me clean up a puddle of puke.

Some of the other sailors told me, "Shit, if somebody's fool enough to want to do my job for me, I'd let 'em." This was, I believe, an empty speech. "Here, sweetheart, I'll take that little wheel off your hands" would never have induced a sailor to give up his work in the wheelhouse. No, all of us guarded the perimeters of our responsibilities jealously. I just had to guard mine a little harder. I grew sore from it, touchy.

It got so that I hated to admit to my own ignorance. I was, for instance, coiling the fire hoses exactly inside out, with the nozzle out of reach. A dumb, dangerous practice. When the *Buck*'s engineer tried to show me the right way, I nearly decked him. "Lady," he said, "seems like you got chips on both shoulders." I guess I did.

But there were big payoffs in going it alone. Once I was doing an awkward waddle across the dock, lugging a crate no bigger than a chicken frying pan. One of my inveterate "helpers," a giant dock boy whose forearms were as big as my thighs, offered to "take that little package" off my hands. "Sure," I said, dumping one hundred and twenty-seven pounds of high-density carbon steel crane reel into his outstretched hands. He dropped it. They kept calling me Wonder Woman. I just about went crazy, living up to that name.

A brief rivalry that went to insane lengths developed between me and Harvey, the engineer.

Crew boats dock, most of the time, stern-to the pier. The captain, on the stern controls, reverses the throttles and literally jams the boat into the dock, at high speed, so as to overcome the river's current. The boat usually bounces once before it settles. The captain keeps the diesel power coming so the current won't wash our boat into the boat alongside.

The two deckhands, in this case me the deckhand and Harvey the engineer, stand one to port, one to starboard, on the open back deck, line loops at the ready, poised to lasso the dock's bitts at just the right moment. We were supposed to be standing well forward of the stern so as not to get popped over into the wheel wash when the boat bounced against the dock. But since Harvey didn't want to see any woman finish the job ahead of him, he inched closer to the stern. Because I didn't want to appear to be a slacker, I inched back too. He inched some more. I inched right alongside. He inched, I inched, and we were standing right at the stern when the *Buck* jammed into the dock. Both of us popped up in the air and were just lucky we landed sprawled on deck instead of in the foot of space between the boat and the dock. We could have been squished, then dropped into the wheel wash and chopped into small pieces by the props.

The second time we pressed our luck, Co-Captain Percy drew an imaginary line across the deck with the toe of his shoe. "You two assholes stay behind this line till we hit dock or I'll shut you both up in the goddam paint locker. You hear?"

But they kept on calling me Wonder Woman. "You ain't scared of nothin', are you?" "Not one in a thousand women could do what you're doin'."

At first I took pride in my nickname. But then I began to chew it over. Something there was not quite right. As if Wonder Woman, somehow more than a woman, was not quite a woman at all?

Even today I have to grit my teeth to keep from biting the tiresome doubters who tell me, "Sure, *you* did it, but no *normal* woman could. Take me [or my wife], for example. I [or she] can't even change a flat tire!" Invariably these skeptics cast sidelong glances at the hard-won muscles of my upper arms, as if to say, "See? You were *born* different."

People who have known me all my life are unbelievers, too. The Lucy they know has always been a weak specimen, hard to picture as Wonder Woman.

At any rate, I was uncomfortable with my elevation to One in a Thousand among Women. If I were seen as Wonder Woman, a mutation, an exception, a freak, then the next woman to work offshore (and I hoped there would soon be many) would have to face the same walls of mockery and doubt that were torturing me now.

I didn't like that nickname at all.

I was a week aboard the *Buck* before my crew stopped offering "help," stopped prevailing on me to make their biscuits and sew patches on their jeans. They could see I was carrying my own weight, doing my own job. The working contract was clear. Then, and only then, did I step out of my blur of stress to make a social contract, too. I was surprised to find that these men were fun to be with, maybe even friends.

I told them I wanted a new nickname. Co-Captain Percy dispensed one on the spot: Lucy Panucchi. P'nootch for short. That was more like it.

12

I noted that David Brinkley looked older, more worn, on the *Buck*'s color Sony than on my black and white back home. Captain Willy, Co-Captain Percy, and I watched Brinkley wince over a film clip of women working aboard a navy repair vessel. Brinkley cautioned us that this departure from maritime tradition was only an experiment.

WILLY: We gotta get P'nootch on board there to show 'em how it's done.

PERCE: (chewing cornpuffs loudly) Shoot us a moon, women.

LUCY: Well, your lives are different now, aren't they? With me aboard?

WILLY: (closing his face) Oh, not so much.

PERCE: (chewing more loudly, ogling me) No, not *much.*

WILLY: Well, one thing, I've gotta remember to kick the door shut when I take a piss.

PERCE: I saw you left it open today. I 'bout closed it for you.

WILLY: Yeah, I forgot once or twice, but I never forgot yet to pull on my pants when I wake up. It's a sight. I zip up, cross the hall, unzip, piss, zip up,

cross the hall, shuck off my pants, and get back
in bed. Some thing.

PERCE: At least you don't have to share a room with her
like Harvey does. He sleeps with his *pants*
on.

WILLY: You know, we're used to just boys will be boys.
Walk out the wheelhouse door, piss over the rail.
Those kinda things. Man-to-man things.

LUCY: I could probably survive a sight like that.

PERCE: Well, we didn't want you thinkin' we're perverts
or nothin'. When I scratch my nuts, I turn thisa-
way.

So there I was, Lucy Panucchi, living and working at peace
in the clubhouse that used to have a sign on the door: No Girlz
Allowed. A full-fledged Buckeye, too. A Buckeye was the
primo thing to be that winter in Cameron.

Cameron is kind of a miniature Morgan City, just a long
smudge of dock. From what I could see of it, nobody lived
there at all unless they lived on the boats. But the deep-water
Calcasieu channel runs through Cameron on its way up to
Lake Charles, so Cameron outdid even Morgan City for boat
traffic. Boats, boats, boats. From one-man oyster skiffs to
oceangoing tankers the size of the *Queen Elizabeth*. Still,
there was only one *Buck*, the king of crew boats.

He'd come out of the shipyard just two months before I
boarded him. He was sleek, hungry looking, painted a crisp
charcoal and oyster, pinstriped in deadly black. A palace
docked alongside the standard oilfield boat shanties, he sat
low in the water, thrumming power in every line. Twenty,
twenty-four knots easy (twenty-eight miles per hour, say),
and even day in, day out that's something of a thrill on open
water.

One hundred miles out in twelve-foot side seas, *Buck*
roared and plunged but never did he heel over like some silly
sailboat, nor did he shudder and creak. The wheelhouse was
quiet enough that the Teac stereo came through faithfully,
even in rough weather.

Buck's wheelhouse was every inch a foxy Lear Jet cockpit. He had an autopilot (a spooky enough thing at first sight, the driverless wheel turning, tic-tic-tic), double-decker radar, a loran position finder, a fancy digital readout Fathometer. I could imagine James Taylor and Carly Simon buying the *Buck* for their place on Martha's Vineyard. A few modifications to the working deck (cargo to canopy) and they'd have a couple of million dollars' worth of yacht.

Buckeyes revved on their boat. More conversation and consternation were lavished on *Buck*'s triple diesels than ever were on the Manhattan Project. Every inch of his sharkish hull was examined daily for the slightest scratch, the faintest stain. Every murmur and belch from his engine room was attended as carefully as the heartbeat of a Rothschild heir. For each line in use on his deck, there was an unblemished spare in the storeroom. Even his spare light bulbs were catalogued and inventoried, his turbo socks rinsed out nightly as if they were precious lace underthings, his zebrawood interior paneling lemon-polished to a soft romantic glow.

Without their *Buck*, the Buckeyes would have been nothing but common oilfield trash. They knew that.

During my first week aboard, I heard the crew grumbling when the *CBM-105*, a boat even newer than *Buck*, tied up alongside. The *CBM* was thirty feet longer, and she sat up high and graceless in the water like a baby battleship. The lines of her hull were banal, clumsy, and she was painted a flat John Deere green. No class at all.

But the Buckeyes had heard some gossip around the fuel dock. The *CBM*'s skipper claimed that his boat, with her superior number of diesels, would have the *Buck* eating her wake. This claim was not to be borne.

"We oughta race 'em," Harvey said.

"But we gotta be cool," Captain Willy cautioned.

"We'll piss on 'em," Harvey said.

"Piss on 'em? Hell, we'll plow their ass under," Percy spat, flattening a mosquito that had dug into his neck. "We'll smoke 'em."

I asked how they could be sure. The *CBM* had four diesels to our three.

"They got four eights. We got three twelves," Harvey said. "And that old sow is heavier, a hundred twenty tons to our ninety-five. Besides, we've got a secret weapon. It just so happens I already took the governor caps off of two of our main engines."

A note about diesel engines: Unlike internal combustion engines, which are limited in rpm's by their very construction, a diesel engine's speed must be regulated by a governor. Otherwise the engine will overrev to white heat, blow up.

Buck's governors sat right on top of the engines, easily accessible. Harvey had, in a moment of stoned mechanical curiosity, removed the governor caps. Thus he could wrench back the governors themselves if the *Buck* needed a sudden, insane burst of speed, as in racing. Not that any boat had challenged the *Buck* until the *CBM* came along.

"I can give her a shot of speed right up the old ass," was how Harvey put it.

He showed me how to pry up the governors with a wrench so they would temporarily disengage. He took me with him to a nearby diesel shop to ask a Caterpillar mechanic how long the engines could run without governors. The mechanic didn't blink at Harvey's question, or refer to a manual, either. "Eleven seconds," he said.

That night I overheard Harvey playing John Wayne with the fuel dock man. If that poky old *CBM* should happen to show her ugly face at the jetties when the *Buck* was comin' through, she'd learn what it was to get smoked.

This was all very serious, you understand.

On the following morning, heading out to the jetties with a clear sky overhead, I stepped out onto the *Buck*'s back deck to pitch our leftovers to the sea gulls. The *CBM* was looming up behind us. I waved a sleepy good morning to its crew, and ducked down the engine room manhole. By the time I'd raced forward to the wheelhouse, Perce and Harvey had joined Willy there with the same news flash. The race was on.

"Stations, then, everybody," Willy giggled, bouncing in his pilot's chair. A nudge from Perce restored Willy's cool. I followed Harvey to the engine room.

We were running at clutch speed, and planning to lie back for the first four seconds (synchronize your watches, everybody) after Willy engaged the throttles to signal that the race was on. Then Harvey and I would pry up the governors and give the engines the goose of unlimited rpm's for eleven seconds. Ten point fifty seconds, really, "just to be on the safe side," Perce had said.

Even at idle speed, the aluminum engine room was loud, loud, loud. In our rush down the ladder, Harvey and I had bypassed the ear protectors. Sound can be as great a hell as heat, you know, snapping brain cells like so many kernels of popcorn. But Harvey seemed unaffected by the mad, clattering roar. He was somewhere else, wherever it is men go in the moments before a heroic competition, biting his lip with an uncharacteristic ferocity.

Strangely enough, the dangers involved in such an undertaking didn't cross my mind. I just hadn't thought it out. A diesel engine, big as a Corvette and many times as dense, might just blow up in my face. As I leaned over the scalding hot turbos and pried up the starboard main's governor, I was taking my limbs, if not my life, in my hands. I just didn't think.

Now. Ten point fifty seconds and counting. I'd bought a new sweatshirt jacket just that morning. I admired its tasteful sky blueness, its woolly-bear comfort. The pockets of that jacket vaporized, disappeared without a trace, when I leaned across the turbos. Even today I treasure that jacket, the only real trophy of the great *Buck* race.

And counting. The diesels sang, screamed, ululated. Their sound so buffeted me that I didn't feel, let alone smell, my pockets burning. I just braced myself on a stanchion and held on to the governor for dear life, watching my watch.

Harvey told me later he'd felt the *Buck* stand on its stern, that he nearly slipped and fell to the deck plates. I didn't remember that part. All my senses were running on overload. What a surge!

As soon as the ten point fifty were up, we flew up the ladder to see if the *Buck* had won. But before we could peek out the rear port, we heard Percy whooping. "We smoked 'em! We fucking God smoked 'em for sausage!"

Behind us, the *CBM* was a gratifyingly puny dot in our wake. We heard its skipper mumbling some soremouthed nonsense over the radio about a next time. We all grinned at one another, thoroughly satisfied. The *Buck* was king. And we were the Buckeyes, in person.

The primo Buckeye, to my mind, was Co-Captain Percy. Perce was what his fellow Buckeyes called The Animal, although they by no means classed him with the animals, their term for rigrats. Perhaps they referred to the way his passions got the best of him from time to time. An instance: When love struck Percy the month before (fully in spite of his marriage to one tough little pack of lunchmeat named Dolly), he got himself what he called "stoned as a clone" and drove over two gas pumps on his way through the façade of the only Laundromat in Cameron. No doubt about it, the man knew how to do it.

Percy was just twenty-seven years old, but a hell of a natural-born coonass boat driver, loony not to a fault but to a high purpose. He couldn't resist doing what he called "pulling their legs off" by "acting unnatural." He told me he and Harvey liked to dance to the disco eight-track in the wheelhouse lounge for an audience of rigrat passengers. "Hugging each other's butts, too," he said. "With our hands." According to him, "All we was doing was acting unnatural. It pulled their legs damn off."

Only after I'd been aboard four days, and then only after we'd smoked what Perce called "roofer" together, did he finally look me square in the eye. Affectionately, too. About as chilling an affection as I imagine a love affair with Hunter Thompson would deliver up.

Then there was Harvey, the *Buck*'s official and uncensored unconscious, only secondarily our engineer, raving at full tilt, bursting with wild plans and emotional noise. He was, for me, an unexpected and encouraging discovery in the Redneck Belt: a genuine Russian Jew. Maybe the onliest little Russian Jew in the offshore oilfields. My joy at finding Harvey among the heathen Christians was what first alerted the Buckeyes to Harvey's alien status. Eyes rolled. Harvey blushed, and stammered.

He was the only son of a controversial right-wing American Army officer, retired now to the status of soldier of fortune, and a Missouri pig farmer besides. So although Harvey must have grown up under an intolerable load of macho expectations, they'd left him unmarked. If anything, Harve was all too human. I sat through a dozen tellings of how a metal shaving flew off the mast and into his eye, and how the nurse held his eye open with long red fingernails.

Unashamed, Harvey told us he nearly blacked out whenever he had to climb the mast to change the anchor light. And when we ran in fog, he admitted he couldn't sleep for fear of drowning. Harvey, God bless him for it, gave voice to our most craven fears matter-of-factly, while the rest of us shrugged that death by drowning never crossed our minds.

All of the Buckeyes were handsome dudes, but Willy had it all. He looked like a rich hippie: long hair balding and graying in a distinguishing pattern at an early age. Wire rims, Calvin Klein jeans, L. L. Bean flannel shirts, Frye boots, a warm little square smile. He could have passed for a compassionate humanist, on looks alone. But where had Willy been in 1968? At an army camp in Georgia, just getting by. The empathy just wasn't there.

Still, Willy said he made it a point to take two Stresstabs a day and keep up with the war news. "I don't want to lose my perspective."

The evening I met Willy he avoided looking at me directly. The next day he sat down next to me and laid his head in my lap. "What I want to do," he said, "is go all over the United States of America with you and look at everything there is." I fell in love with Willy.

Every time he and I rubbed elbows, we lingered over the electric tingle of our touching, the kind you could measure with a circuit tester. And when we leaned over the radar simultaneously, we fogged it with our hot breath. We made stoned plans for a cross-country tour in the spring. We watched over each other, as I remember it, gently delaying the first move.

I spent one whole afternoon watching his bare foot swing loose as he napped, sitting upright in the co-pilot's chair. It was raining, and the *Buck* rolled gently on the gray Gulf. We made dinner for the crew that night, whipping up biscuits together, teasing one another about playing house. I'd been wounded not long before, and I felt a little bit wary, but not very.

Lonely, I was lonely, lonely, lonely, wretched, aching lonely, an alien and a target among the rigrats and sailors who dealt me their dead-serious True Love games from crooked decks. So many woman-starved men all seedy and needy in their loneliness, wanting just to touch me, just to prove they were men—I'd never touched any at all. I had so rarely been without a man. By the time Willy and I exchanged electrical charges, I'd reached critical mass.

Still, hadn't I sworn I'd never again get intimate on such a spur of the moment? Hadn't I vowed never again to slide crabwise into a love trap? And Willy was my captain, my boss. At least Willy wasn't married, or even attached. And he didn't seem to mind when Perce teased him about having a crush on a deckhand. But, but, but.

Percy, my sometime confidant, cut short my precopulatory quibblings. "Shit, woman. If I was you, I'd grab all the good fuckin' I could get. Screw me a blue streak up the middle of the Calcasieu channel. I sure wouldn't piss around worryin' about it, me."

I decided, sure, I'll go ahead and get next to Willy. It couldn't hurt. It might heal.

And it was fine when we hugged two by two by the TV, feeling the warmth of his fine-boned body, basking in his shy smile. Going beyond that, going down to Willy's stateroom, was the mistake.

First he turned off the lights, locked and double-checked the door. Then he removed his clothes and arranged himself between the sheets. A little too businesslike for my tastes. But I slid in beside him and snaked an arm around his neck. He kissed me on the corner of my mouth, adjusted to achieve full mouth contact, spread my legs with one hairy knee, and made as if to mount.

Hey, hold on here, I thought. In two point three minutes, according to Dr. Kinsey's best estimate, It will be Over. That wasn't what I wanted at all.

I flinched out from under Willy's standard approach pattern. Remembering my last brutal love encounter, I sat up, stubborn in my soul. I am present here, a real woman, and my feelings must be reckoned with. I like to touch, touch and be touched and lose myself in that. I am not here to procreate, but to make love. "Hey, Willy? We've got some time tonight. How about if we get tender and silly. Let's have fun." My voice sounded both crackly and childlike. My heart was in my throat.

Silent moments and half-hearted breast brushings later, he again made as if to mount. I stayed him with a mild gesture, but a brave one. It's scary to stick your neck out, trying for intimacy. "Willy, could we get into some playfulness here? Some, you know, erotic sex? Like this?" I demonstrated.

"Well, sure, whatever you want." Accent you.

When, a full minute later, he moved doggedly to the colonial posture, I gave in and spread. Let it be. I hadn't learned, then, that I could simply halt the process, get up and go. No, I thought, that would be *awkward.* I stayed.

Once Willy got me properly stuck, he buried his poor, anxious bald head in my neck and burrowed on down for the two-point-three-minute mile, hands splayed out on either side of me in pushup position, pelvis hammering like a moth at the window.

Lucy the bedtime Wonder Woman clicked on. I writhed and moaned heroically, such a little liar with a heart as empty as a tomb. I'd sworn I'd never do this again. But here I was betraying myself, betraying him, too.

His response to my well-camouflaged shame and resentment? Get ready, the male lead is about to deliver his best line: "Ohhhh, Luuucy . . . where have you been all my life?"

That's what he said. This dreadful disconnected pumping was Willy's idea of fun. At least of fun done correctly. Unless he was pretending, too. That's possible.

Inevitably, he bumped and buckled, exploded with a brief grunt, collapsed. For a moment there I loathed the man completely.

Then my loathing took an inward turn. So this was the new honesty I'd vowed to live by in matters of love. I had liked this fellow very much just a few minutes before. Enough now to stick my neck out again? Reveal my own softness, my ticklish fantasies, my fears? I wished he were not as scared as I was. Maybe he was more scared? Most likely he would say he was not scared at all. "Scared of what?"

Propped up on one elbow, brushing the hair out of my eyes, I blurted out some heartfelt message about how frightened I was of sex, of intimacy, of genuine touching. But how I couldn't see what else sex was for. I told him I was trying to relax and play and let more of myself show and—

"Let's talk about it in the morning, OK?"

"OK," I said, stung, brought up short.

I feel sad for those two people lying curled up not sleeping, alone all night in that narrow bunk of a boat far out in the Gulf. Everything they need is right there naked and trembling, just beyond the wall of self-consciousness. Everything it takes to get down. Everything it takes to get to heaven. But there they lie, immobile as two skinned rabbits in a deep freeze. I can't know what passed through Willy's mind that night, but I do know what was on mine. It was a question I'd heard years before from a traveling Hindu sage. The question: What is it that keeps us from the Beloved?

13

Willy and I kept a businesslike distance for many weeks after our night in the Alone Zone. On long watches, with just the two of us in the wheelhouse, we didn't even talk. When we passed each other in the narrow aisles of the lounge, not even our rain slickers touched. I hadn't stopped wanting him. I thought we'd had some minor kind of misunderstanding, something we could talk out. I thought at least that Willy was worth retraining in the arts of love.

I remember feeling a tug of regret when I overheard his distinctive giggle. And I'd have to swallow a sigh when he climbed the ladder of the stern controls ahead of me. I missed being able to touch him, missed being held. This time I was the rebuffed True Lover, hurt and puzzled as the rest of them. Angry, too.

Love, I told myself, is the least of my worries in the winter Gulf. My top priority was, or should have been, sleep. A month-long fog blanketed the coast, grounding the helicopters and intensifying the demand for speedboats like the *Buck*. We were running hard twenty hours out of twenty-four, sleeping three of those hours if we were lucky, and sleeping not at the dock, in peace, because we had chores to do at dock, but out on the choppy Gulf.

Sleep. I could have cried for sleep. I came to hate the company radio and the dispatcher's macho hard-mouth singsong, calling the *Buck* at all hours. After fighting to stay alert and awake through a long watch, I'd no sooner lie down to rest than I'd hear the radio fizz and crackle again: "Craink her up, fellers. Time to head out."

There is a twilight zone of frustrated exhaustion where the boundary between waking and sleeping crumbles. Breaking point is so long past that it blurs in memory. Nodding over a wheel watch, I'd hear a familiar voice, brimming with distress, crying Mama-mama-mama! I'd jolt up, wide awake, knowing it was my own voice I'd heard crying aloud. Occasionally the voice was more succinct: Help!

Then there was the supreme of minor irritations, sleeping in my clothes. I slept in my clothes all that foggy month. The watch that I wore every hour of the day threatened to grow into my wrist. My waistline blistered under weeks of painful binding in damp jeans. My shins and calves were chafed raw by saltwater-wet rolled-up jeans lived in, slept in too long. I developed curious sore spots in my armpits and at the fold behind my left knee. Changing to dry clothes wasn't worth the trouble; within the hour I'd be stepping out on deck again where icy gray waves slapped my pant legs and filled my boots.

Harvey Hertzberg, my bunkroom mate, had slept so long in his clothes that he was unable to sleep at all unless he was engulfed in his clothing, all his clothing at once. We called his bunk The Garment District. The engine room, where we hung our wet clothes on wash lines strung between the hot diesels, was The Garment District Annex. Leaky laundry piled up in the passageways, in the showers, on the galley bench. The Buckeyes' usual camaraderie gave way to snappishness. "Who took my goddam jeans off the line? They ain't even dry yet, goddammit." "Who's the fucker used up all the dry towels?"

Sleep. All we needed was a little sleep.

I got so tired of trying and failing to get some sleep that

I gave up sleeping altogether. I took up the practice of falling out instead. I could fall out anywhere at all with my hat on and my boots for a pillow. In heavy seas I'd grip an overhead beam with one hand to cushion the impact of three twelve-cylinder diesels driving aluminum hull through eight-foot waves bow first. If I didn't sleep, I at least lost consciousness. Sometimes I came to in midair, as when the *Buck* plunged into the trough behind a monster wave. Sometimes I came to on the floor, bruised and cursing.

Soon I joined the ranks of master sleepers who can practice their art in steel foundries and high-speed electric drill testing labs. I became a wizard of snooze who could catch thirty winks standing upright in freezing rain.

I seldom fell out in my own bed. For one thing, my bunk shared a wall with the engine room. Sleep to the engines' howling topspeed vibration wrung me out. As often as possible, I sacked out abovedecks in the passenger lounge, twisted around the rigid arms of the theater-style chairs.

Driven belowdecks by a cargo of noisy rigrats in the lounge, I'd fall out on the galley bench, bracing my arms on the opposite walls of that narrow space, presumably to keep them from closing in on me. But who can sleep in a ready-made coffin? The *Buck* might ram a tanker in the fog, for instance, or roll over with a vicious side sea. I imagined stairs upended, a wall of angry water behind the engine room hatch. The *Buck* owned copies of *The Titanic* and *The Poseidon Adventure*. All of us had read them.

Drowning is only one of the panoply of specters that can kill off a sailor's sleep. Every time we hit the dock, I carried our garbage to the dock's trash bins, The Rat Zoo. I'd seen B-movie rats who rear on their hind legs and show their yellow teeth. But these real-life rats, a couple of hundred strong, panicked when I came near, stampeding in a bubbling flux of furry humpings. Their naked feet made a soft *thipping* sound on the rotting pier.

I tell this to demonstrate that if you harbored a secret fear, a private horror, you would sooner or later meet it face to face in the sailing life. And sooner or later that fear would climb into bed with you to steal your sleep.

If you feared heights, there were helicopter flights, basket lifts to the rigs, masts to be climbed. For claustrophobes, the boats offered airless mud tanks, the narrow cells of staterooms below the waterline, the suffocating din of the engine room, the lonely horrors of the anchor hold. If emptiness and silence chilled your heart, you'd find plenty of both on deck in the winter sea, where gray sky met gray water seamlessly. When you came to think about it, which you surely would, eerie beings slithered awake nightlong on the other side of the hull from your bunk. Sharks were present in force. Everyone knew of a sailor who'd been struck by lightning, or lost overboard, burned alive, blinded by acid, or crushed by some giant falling object. We had time to ponder these things, at sea.

The special horror that haunted me derived from a story Guste had told, about a barge that overturned near the mouth of the Mississippi. Guste's boat, the nearest, stood by. A Coast Guard commander came aboard and strained alongside Guste for eighteen hours to right the sinking barge. Trying this, trying that, while time slipped by.

Guste's radio relayed the screams and pleadings of the barge's crew trapped below with four inches of air, no more, according to the one man who managed to escape.

When the captives' air began to go stale, the Coast Guard man switched the radio off.

It was late afternoon at the dingiest hour of the day when we brought the *Buck* into dock. We'd been a day and a night on the Gulf moving rigrats from one platform to another. Our dispatcher hopped aboard with his usual proprietary smugness. We Buckeyes might live aboard the king of boats, but the dispatcher is the pope who sends the king running.

"Crew change on Cactus 102," he smirked, glad bearer of bad news. "Better gas up and move out."

I bit back my frustration until he was out of earshot. "Go out again? We just got back! We can't go out now . . ."

Harvey just grinned, showing the spirit. "As long as we've got drugs, we can do anything."

Harvey took on fuel while I gave the *Buck* a quick freshwater rinse and filled up the water tanks. The chilly mist of that activity refreshed me. Sore and exhausted as we were, it was true: We were the Buckeyes. We could do anything. We certainly had the drugs.

Harve and I smoked a couple of jays on deck, overlooking the evening parade of shrimp boats that buzzed into the channel from the jetties. Time to go out.

Inside the passenger lounge, we found Willy sprawled asleep, Perce absorbed in a tantalizing TV drama: a deadly cobra lay in ambush for juicy Susan Anton. Perce rooted for the cobra. "Go on, bite her! Sting her! Wherever you hit her, I'll come behind and suck the poison out myself. Willy, just look at the hockey box on this 'un."

Harvey shook Willy awake and offered him a cool cup of thick Louisiana coffee. Willy laced it with a shot of Old Overholt. Time to go out.

Willy and Harvey took up their vigil in the wheelhouse while Percy and I put our feet up in the lounge. "Let me tell you about the night me and Harvey rode the White Buffalo," he said.

There followed an all too graphic account of a seedy seagoing *ménage à trois*. "That old troll there, we picked her up on the dock. She didn't have no hockey box. She had an old giant-size sandbox. I come in the galley and I seen Harvey down between her legs, lickin' her chops, you know, and me I hadn't never done nothin' like . . . nothin' I could call . . . *bizarre sex.*

"But I figure here's here, now's now, if you're ever gonna get some *bizarre sex,* this is gonna be it tonight. So I climb up on her big ol' chest. Harvey asks me what the hell am I doin', and I say 'Just you watch 'cause I sure am doin' it.' 'Gobble my goober, you animal,' I said to her."

Hours later and the warm wind from the south shifted to punish us, cold and implacable out of the north. We'd just taken on the last of a load of rigrats, weary men already disgruntled at having their helicopter crew change delayed two days. Rigrats, as a rule, hate riding the boats, and take a kind of inverse macho pride in getting sick on them.

As we turned to head home, against the seas, the *Buck* began what is called pounding, a term that refers to the smacking a boat takes when it hurtles head on into steep waves. The *Buck* lurched, crashed, rose dizzy over the seas, and fell hard into sickeningly deep troughs. I passed out the seasickness bags. Percy took over the wheel. Willy and Harve, the *Buck*'s weaker stomachs, numbed their nerves with a bottle of Formula 44 Effective Strength and half a pint of George Dickel, then stumbled off to their bunks.

Now the *Buck* was a floating cattle car. Rigrats hid their heads in plastic bags: *bleeaachk burrrrh bleahhhhh BARRRRRAHH.* Rigrats lay right out there in the aisles, shuddering sick with heaves that racked their sore innards, every face ashen and desperate. Six hours of this from the rig Cactus 102 to Cameron.

Any serious captain would have clamped his jaw down tight and held our course in silent fury. I was pretty disgusted myself; it was me who'd be dredging out the lounge when we hit dock and the rigrats debarked. But Perce, our master of malicious mischief, rose above it all. To every wave of cattle calls and moans, he jeered back, "Pork chops! *Greeeeasy* pork chops with pussy juice on 'em!"

"Man, you're *cold!*" a rigrat groaned in response.

"Hey, I ain't cold, I just ain't no damn pussy. I wouldn't be comin' out on this water if I couldn't take it. Grillllled sausage! Greasy lumps with boogers on 'em! Hoo-hoooo!"

Later that night Harvey took over my watch so I could fall out. But I was lying awake down in my bunk, battling avoirdupois for equilibrium, when the grease-drenched odor of frying bacon slithered under my door and tweaked my nose.

A temptation to quick give up the contents of my stomach. A temptation to holler at Percy that his torture of the rigrats was going too far. But, hell, like Willy often said, you just can't let things get to you. So I got up again and joined Perce for breakfast. Good bacon.

A theory of *mal de mer:* Seasickness is, I believe, a problem of innard identification. Innards and outards traveling at differing velocities, in unpredictable directions. Identifying with one's outards in the usual upright anthropoid manner puts one out of synch with Mother Sea. Movement in the *hara,* the solar plexus, that most innard of innards, follows perfectly the movement of the great waters. Identifying with the *hara* permits one to float comfortably on the here and now. Cold air helps, too.

Later that morning, dazed and exhausted, I did what Willy called "a fool thing." We'd stopped to discharge our suffering passengers, and were bound for the fuel dock again. I jumped down to dock to free our stern line from a high bitt (absolutely fullest concentration required to make a successful leap from revving boat to slippery pilings) and then turned back to see the *Buck* drifting away. I didn't think to stand my ground and wait for the *Buck* to back up for me.

Instead, I saw my missing-the-boat nightmare coming true. They're leaving without me! I didn't think at all. I leapt, flying in a great dreamlike leap ten feet or more over the rage of Calcasieu River current and the *Buck*'s deadly propwash, just scrab-grabbing on to the *Buck*'s bumper tires with my fingertips.

I didn't want to miss my boat.

It had been two weeks since we'd had more than a couple of hours on shore. The dispatcher was ready to send us out again one night but we decided to fake some engine trouble so we'd have time enough to do our laundry. Since Percy had destroyed Cameron's one and only Laundromat the month before, we headed for a rut-roaded backwoods suburb.

"Guess we'll be washing our clothes over in Niggertown," Harvey sighed.

You fucking racist, I thought.

"Niggertown's got a Laundromat," he said.

And sure enough it did. Or used to. Apparently the dryers had once set the roof on fire and burned it down to its rafters. The place was open twenty-four hours a day because there were no doors and never had been. Neither were there change and pop machines, nor a laundry warden on duty to chase out the neighborhood dogs that coupled between the canted, abandoned machines. There was not even a floor, and never had been one of those either.

The washers were a monument to someone's basic indifference to the virtues of maintenance. Most stood open to the night sky half full of murky rainwater. All but two of the washers with lids in place were as full of water, still brown water with moldy blue eyes growing on it.

The dryers stood along one wall, blackened and twisted by what must have been a very hot fire some months in the past. We'd have to dry our clothes in the engine room again.

But since two of the washers accepted our coins and dispensed cold water into their tubs, the four of us lined up to launder our jeans and T-shirts, horrified and disgusted or no.

I noticed that some local Kilroys had left their marks on one of the remaining walls. And while most of the graffiti told of petty crushes probably long since forgotten, one scrawl spoke in a more philosophical and persuasive voice: CORNBREAD LOVES WOMENS AND YOU.

I would learn more that night about love.

Percy's extramarital girlfriend, Kathy, drove up out of the night in her sharp new Mercury Cougar. But Perce was hoping to avoid her now. His wife had laid down the law. So the Buckeyes asked me to take a ride around town with Kathy, to catch some fresh air and keep Kathy's mind off her man.

Kathy hadn't invited me, but she didn't blink when I lowered myself into her crushed Dacron velvet upholstery. In fact, she acted as if we were resuming a conversation just briefly interrupted. I'd never met the woman in my life.

She picked up in midsentence: ". . . and Percy means more to me than inny man ever did. I just feel really good with him, you know? They tole me you're Willy's girl. I allys liked Willy. He's mellow. Percy ain't mellow, but you know, he just makes me feel so good. I heard his wife found out I come on the *Buck*. Now he don't want me comin' around, he's afraid she'll come on and see me. That bitch. She really is. I'd scratch her damn eyes out if I saw her right now. I love Percy."

For as long as Kathy spun on mindlessly, I was relieved of my obligation to the conversation. I noticed how her hard little blond face seemed pressed under glass, immobile. Her lips barely moved, but still issued a stream of flat noise. ". . . and I been goin' out since I was fifteen, so I know what I need to know about men. Percy is different. He knows everything, he just ain't sayin'. He just listens and looooks at you that way. He's more to me than inny man ever was, I mean it."

"How old are you, Kathy?"

"Be eighteen this summer. I'm a Leo, same as Percy. You know what I mean about him? He's just the way he is . . ."

While Kathy read out the tangled history of her True Love, I sank back into my seat to more fully appreciate her driving. Smooth and practiced, she negotiated downhill turns into washboard back streets and wheeled out onto highways still under construction without appearing to be aware that she was driving at all. Kathy in another context could be the next Janet Guthrie.

Her eyes flickered over an occasional barroom parking lot. Checking, checking.

I asked where we were going.

"Oh, I got this kind of like a route I drive around town, looking for different people. I find them by their cars, where their cars are parked."

"What people?"

"Oh, captains mostly. There's this one captain on the *Puma*. He wants to take me with him to Sibley."

"Sibley?"

"Yeah, Sibley Australya, I think it is. He said it was a real famous place in the world, like Paris or somethin'. He's goin' over there with some big oil comp'ny, drive their boat for them. You tell Percy I said I was goin' on the *Puma* if he keeps on with Dolly. Why should I stay here in this old hole when I could be goin' with different captains to Sibley and places like that?

"What Percy don't understand about me is I would do anything for love. Anything, you name it and I would do it for love. I would go to Sibley. I would go further than that. I would kill for love."

It was deep in February, the cold heart of winter. I'd spent more than a month on the *Buck*, no longer a freak and an outcast, but happy with my crew, at home at last. Which is not to say that I had stopped playing Wonder Woman. Standards of performance, once set, are hard to abandon. Particularly on the *Buck*. All of us babied the *Buck*. All of us worked too hard.

Willy left for a few days ashore, squeezing my hand meaningfully on his way out. He returned, he was bearing gifts: a pair of dick-nosed eyeglasses for Percy ("Wait'll the rigrats see this shit. Freak 'em out!"), a pound of Texas homegrown dope for general use, and for me a nervous embrace. Willy even looked me straight in the eye, with a smile, for the first time in weeks. My heart fluttered like a silly, helpless bird. True Love's reward.

The two of us shared a wheel watch that afternoon, and then sneaked down into his bunk when Harve and Perce came on watch. I figured Willy had maybe thought this whole thing over in his time off, was maybe ready to get close. That may have been so.

He did tremble when he held me and kissed my ear, both of us sitting naked and shy on the edge of the bunk. But when he hugged me again, he spilled over. I felt him cringe. I made the mistake of saying it didn't matter. Obviously it did. We spent another sleepless night in the Alone Zone. The next day he turned ugly.

We were docked in Intracoastal City that morning. Normally we'd never have gone there, but this was some oilfield fluke. The *Buck* dropped me off at the grocery store to do some quick shopping. I was feeling down, I remember, dragging my heels in the road, puzzling over what was going wrong between Willy and me, when I caught sight of a turnip-shaped man in a gaudy jumpsuit. It was Guste. We roared into each other's arms and danced a bear hug. I hadn't realized how much I missed him. He hadn't known where to find me. We leaked tears and laughed all at once.

Guste introduced me to his new cook, a bright Florida girl with big green eyes. "Dis de one I tole you about, dis de one," Guste kept telling her. "She sho' cut dem han's a new asshole, her."

I invited Guste and his cook to come visit on the *Buck,* and we picked up Cupp and Fred Fatigué on the way there. Neither Cupp nor Fred offered to hug me, but they couldn't seem to stop smiling at me, either. My Buckeyes liked me, but not in this warm way. I hadn't felt so welcome and so loved in a long time. We climbed aboard the *Buck.*

While I gave my friends a tour of my glamorous new home, the Buckeyes lounged deadpan in the wheelhouse. The uneasiness between the two crews surprised and confounded me. I couldn't imagine, and still can't, what was going on below the surface of this chance meeting. Some maybe southern thing, a class difference or a race thing on Fred's account? I didn't know. I still don't know.

When my old friends had gone again Willy asked me, "S'poze you could get me fixed up with that girl cook from your old boat?"

I felt that wallop in the solar plexus. Was he serious? Jealous? Crazy? What?

I recovered my breath and asked just what did he mean by that crack? Was he trying to hurt my feelings?

Willy's answer was oblique, another slam. "Maybe I'll go down to your bunk tonight and get me some more of what you give away all over the docks. Bet you don't never run out of it, slut."

I stammered, frozen to the spot. All I could do was wince and stutter, "Wait. Wait, wait, just wait a minute, Willy," as if I could halt this craziness, run the film back and splice in a different scene.

"Shut up and make me some coffee, slut."

As many times as I've relived that moment, I still don't have a grasp of it. I simply don't know, probably never will know, what was happening in Willy. Gentle Willy, uptight Willy, my friend if not my lover. I just backed off and kept to myself. All that day the Buckeyes, led by Willy, teased me, ragged me. The ragging got more vicious every hour and I just couldn't figure it out. These men had been my friends.

On our way out of dock that night I called the *Pride* on the *Buck*'s radio. To say goodbye to my only trustworthy links to sanity. I couldn't raise the *Pride*. Repeat call. "This is Lucy on the *Buck* calling the *Harbor Pride*. Come in, Goose . . ."

Willy leapt into the wheelhouse, snatched the mike out of my hand. "You're just fooling yourself, you dumb slut. Those people don't give a shit about you."

Before I knew my fist had left my pocket, it was smashing into the bridge of Willy's nose. Harve and Percy jumped out of their seats and grabbed Willy before he could bash me back. I had by then hunched myself into a corner, horrified, cringing.

"You kin pack your shit and hit the dock when we get back to Cameron," Willy snarled.

I spent the intervening hours in the galley, waiting for Harvey or Perce or even Willy to come talk to me. When I thought about approaching them, in the wheelhouse, my feet refused to take me there. I was scared and giddy, disoriented, ashamed. I had hit somebody. That made me a hitter, a bad bad girl, not Wonder Woman at all.

But somewhere in there I stopped shivering with shame

long enough to remember how good it had felt to let loose my anger. Good and satisfying. I'd struck back.

Harvey appeared once on a coffee run. He smiled a tight smile, embarrassed. "That was pretty heavy, Luce. Pret-ty heavy."

I shrugged with a pretended nonchalance. Maybe I was still a little shaky about it, but I wasn't really sorry anymore.

Harvey was right, though. Striking your captain, even if he's your lover, is a merchant marine capital offense. Percy's brother spent three years in a federal pen for suggesting that his captain stick something up his ass. I had gone further, into violence. I'd be lucky if Willy didn't press charges, but when I thought about it, I knew he wouldn't. I couldn't imagine him standing up in front of a Coast Guard tribunal saying, "This woman hit me! With her fist!" Never.

Harvey ducked away, back to the boys in the wheelhouse, leaving me alone with the monotonous roar of *Buck*'s diesels. Homeless again. Maybe, I thought, I never will find a home out here.

14

Our dock's dispatcher, who had some old grudge against Willy anyway, offered to "flatten his ass" for me. He offered me a job on his dock, too, running a forklift truck. That might have been fun, but already I was feeling a little landsick. If the dispatcher could help me find another boat job, that would be enough. With just two phone calls, the dispatcher got me a place on another crew boat, the *Empress*. He even arranged a helicopter ride to take me to her dock in Dulac, Louisiana. Dulac was all the way back east, toward Mississippi.

Waiting for the helicopter, I met another onliest little woman in the oilfields. Her name was Bonnie, she cooked on a rig, and I was the first female she'd seen in a month. "Gits so you fergit what a woman even looks like," she said.

She wanted to know how I'd stood it, working from the Cameron docks. "Honey, they got rats this big in Cameron, and nothing they like better to eat than human flaisch. These here little crabs, the ones 'at jump out at you with their eyes up on little sticks-like? They're bad enough. Make my skin creep. But ra-yats!"

I wasn't looking forward to my first helicopter ride, but when the chopper came, Bonnie's fears put my own in perspective. "You mean they ain't gonna shut that thaing off 'fore we get in it?"

No, I didn't think so. At least they never did in the movies.

"Well, then, I ain't gettin' in it atall."

The pilot of that chopper, faceless behind helmet and goggles, offered me a seat up front in the plastic bubble, the co-pilot's seat on the edge of blue nothingness. I accepted. There were some advantages to being a female in the oilfields.

As the chopper lifted us up from the docks, I saw Bonnie growing smaller as she hoisted her seabag onto her back and trudged down the muddy road.

Once we were level with the clouds, I remembered I'd planned on suffering from my old favorite phobia, vertigo. But the view was too compelling, the chopper controls too fascinating. This was the ultimate thrill ride in the great oil-field amusement park.

Only the noisy *whup-whup-whup* of the rotor interfered with my joy. The pilot remedied that with a pair of earphones I could switch from the chatter of air traffic to the pilot's personal cassette system. "Running on Empty" pumped into my ears, shutting out the distracting racket. The only improvement this experience could stand was a couple of tokes of marijuana. I rolled a joint of Willy's homegrown and shared it with the pilot.

We were flying over the Gulf by then, sailing through wispy white clouds with the first bright sun of the week at our backs. Our most recent storm had wrung the skies clear and moved on.

Below us the gray shallows shaded into turquoise seas. Erector-set cities of rigs, platforms, well caps, appeared. Toy boats buzzed beneath the rigs and between them on the errands I knew only too well. It was great to be, for once, above it all.

The pilot took us higher, toward cloud banks that looked like great frozen oceans. When we flew into the first bank I flinched, expecting impact. But only the newly muffled *whup-*

ping of the rotor and the dimming of sunshine to twilight marked our entrance to the cloud kingdom. Inside the seeming solidity of the banks, great sunstruck caverns opened to us, vistas within vistas.

The pilot's voice, soft and dreamy, broke into Jackson Browne's "Just a Little Bit Longer" to ask me, "Have you ever put your hand in a cloud?"

What a question.

He opened the vent on his side. I opened mine and stretched my fingers outside, tentatively. For all I knew, this was some oilfield frostbite joke.

The cloud was only cool, as wet as running water. But when I withdrew my hand from it, thinking to taste fresh rain, my skin was perfectly dry.

I came to earth half a muddy mile from the *Empress,* my new boat. The personnel man for the company that owned her was waiting for me. "You always carry this typewriter with you?" he complained as he loaded it into his car.

When we reached the *Empress*'s dock, I read the boats' names on the sterns backed up to the piers. Which one would be my *Empress?*

"Thar she blows," the personnel man cried out with false heartiness. There she blew indeed, carelessly tied to a rotten pier, listing to starboard, crumbling into rust right before my eyes. She was just a small boat, I'd been prepared for that much. But I hadn't expected this . . . this . . . *African Queen.*

Her lines were shabby, dumped rather than coiled on the bitts. Her paint was peeling, or had long since peeled. The rust on her hull was so old it formed bubbles and slabs as thick as shale deposits. She had, the personnel man told me, just returned from four years' duty in the Central American oilfields. I could smell it on her, an air of don't-give-a-damn tropical decay. I could smell danger, too. A boat in this ruinous condition was no match for hard weather. "We'll be putting her in shape at drydock pretty soon now," the company man told me, lying through his teeth. Then he put my seabags on the dock, introduced me to the deckhand, Mighty Mouse, wished me luck, and buzzed away in his shiny Camaro. Glad to be away, too, before I saw the worst. I climbed aboard.

Every boat on the water takes some getting used to. Each hull responds to the sea with its own distinctive roll; every wheel has its quirks. But that third boat of mine, that *Empress*, went beyond quirky, beyond erratic, all the way to recalcitrant, treacherous, loony. She was stove in on all sides, patched together, and sucking up seawater at every seam.

Her starboard rail, busted loose at deck level, waggled over the water like a limp distress signal. And according to Mighty Mouse, the *Empress*'s list to starboard was more or less permanent. Her anchor winch, on the forward deck, was rusted solid. Up on the roof of the wheelhouse, her spotlights swung with the roll of the tide. Mighty Mouse, a mild blond Arkansas giant, told me that the *Empress*'s port engine was "bad off," twenty-nine thousand hours without an overhaul. Good God.

Her house, her indoors, was in no shipper shape. Stained and ragged curtains drooped over the windows of her passenger lounge. The door to her abovedecks head was nailed shut, with OUT OF ORDOR scrawled in flow-pen on it. Up in the wheelhouse, I found a blur the size of a silver dollar in the center of her radar screen. "Radar don't work half the time anyway," Mouse reassured me, as if two wrongs made a right. A pair of fuzzy dice hung from the spotlight controls. The navigation charts were not rolled and stowed but folded and smooshed into a damp corner of the wheelhouse floor. Just standing at dock like she was, the *Empress* leaked.

Our tour belowdecks was no more heartening. Mouse introduced me to the one functioning toilet, and cautioned that I could urinate there, but never defecate. "Use one of these plastic bags," he said.

A bag?

"She won't flush nothin' solid."

The freshwater tank had literally come apart at the seams;

when the *Empress* came ashore, the crew filled gallon jugs at the dock tap and lined her cabinets with them. Mighty Mouse told me I could pour two of them over my head for a "nice little shower."

Since this was a four-man-crew crew boat, there was no cook in residence. (It's the Catch-22 of the oilfield merchant marine. With a four-man crew, you need not hire a cook. But once you add a cook and bring the crew to five, you're justified in having a cook.)

The galley was positively sinister, showing the typical transient's lack of concern for cleanliness and then some and then some. The freezer was stocked with pork chops and beef roasts, the cabinets with Shake 'n Bake, Rice Krispies, canned white beans, and roach traps. "But now that you're here . . ." Mighty Mouse began. "I'm no cook," I lied. "Couldn't cook if I tried. I was hired on for deck, like you." He took that well enough.

We tiptoed back behind the galley, past the single stateroom where the captain was sleeping. Behind that were what looked like Pullman berths. "This here's where we sleep," Mouse told me. Sam, the engineer, woke from his nap and slid down out of his berth to greet me.

"So this is it," I said.

"Yeah, this is it. Hellhole, ain't it? You a good cook?"

"Can't cook a lick," I lied again. "Burn everything I touch. But it looks like this boat could use another deckhand, just to kind of clean it up."

"Don't bother," Sam said. "Rigrats just come back on and puke all over it again."

The lights flickered while we talked. What was that? "Just the generator," Sam explained. Our number one generator was out, permanently out. The electrical power furnished by the backup generator surged and flickered ominously. "But if the backup generator goes out, we'd be helpless, wouldn't we?" I asked.

"Company don't have the budget to fix it yet," Sam told me. As I unpacked my bags into the berth above Mouse's, I heard the whole story. I was working for a company that was just getting off the mark in its race for black gold, running on a shoestring "temporarily."

"I bought the groceries myself two weeks ago," Mouse complained. "They ain't paid me back yet."

"Least we get our paychecks on time," Sam said.

"So far," Mouse sighed.

Sam went on to itemize further troubles aboard the *Empress.* Our safety vests did not buckle, and every fire hose on board was cracked at the seams. Even the *Empress*'s superstructure was unstable. "She was just made that way," Sam guessed. With twelve feet of empty hull (called The Void) under the *Empress*'s galley deck, topheaviness was the inevitable and unseaworthy result. In a side sea, she simply heeled over and sailed along singing on her bulwarks until the pilot, or the sea, wrestled her upright again.

"Captain don't even have a license," Mouse whispered. "He tole me not to tell anybody, but the company don't care anyhow, long as he gets it before the Coast Guard inspection."

And when would that be?

"I think they said August."

It was still February.

Sam and Mighty Mouse yawned and climbed back into their bunks. The *Empress* was first up for the next offshore run, and it would be a long one. I spent the next hour on my own, peering into hatches, putting charts in order, coiling the lines smartly around the rusty bitts, airing out mops. This *Empress* was my boat now, and it behooved me to make friends with her, but she sure was a cruel comedown from the *Buck.*

It was this *Empress,* this boat I'd hesitate to call a craft, that the four of us rode out into the worst winter storm of this century. I'd been aboard her nearly a week by then.

Our captain, Corey, was beyond logic into Benzedrine that night, having put in fifty-six waking hours at the wheel hunched over our defective radar, navigating in fog. The helicopters hadn't left our dock in days, thanks to the unremitting winter fog and rain. So the crippled *Empress* took on their duty, making double and triple runs on the Gulf. Without the light machinery and supplies we ferried offshore with the crews, a rig might have to shut down. Shutting down a drilling rig can cost a hundred thousand dollars a day. We'd be running offshore again soon, earning our keep.

Captain Corey should have been sleeping while we were briefly in dock, getting ready for the next run. Instead he was speed-tripping, wired to the ears. He went ashore to invite his intended, whom he called Big Bettie the Tiger, to make our next run with us, just for fun.

In his absence I should have been tidying up the wheelhouse, or filling one-gallon jugs at the dock spigot. This had been a hard week for me. A whole new dock crew to train out of being "helpful." And on this sleazy boat, Wonder Woman had her work cut out for her. I spent one whole day scraping sludge out of the oven, another day painting our only working head. I should have been Wonder Womaning right then, but I was just too tired to care. So I leaned my feverish head on the cool steel galley wall and rested my ears in the awesome silence of our rattletrap *Empress* at dock with her diesels shut down. I'd been awake most of that fifty-six hours, too.

And then the captain's woman appeared, feet first as she stepped down the perforated steel stairway, backlit with fog-diffused blue daylight. Her tiny trotters, encased in the canvas proletariat approximation of ballet slippers, belied her marvelous bulk. But, oh, here it came.

Unwittingly the crew formed a receiving line, all eyes upturned in awe of her size. Our captain, a West Virginia gentleman, played proud escort to her entrance. Now *this*, his posture implied, was a *real woman*. We gawked. What wonder could she be if not Last Princess of the Samoas?

But no, her black windbreaker said she was Big Bettie, a Pisces, a Tiger too.

She didn't have much to say for herself. Her eyes were for her prize, her captain, her diminutive red-eyed zoned-out groom-to-be. She tittered in his ear; he directed her to the head.

In her absence, Captain Corey said he'd be hustling her off to the courthouse to make her his fifth wife as soon as we returned from our run.

"My fifth, but not my last," he whispered. "Promised my brother on his deathbed, he'd had him five wives, I'd go him one better."

Did Bettie know that?

"Hay-ell no. Think I'm crazy? They don't call her the Tiger for nothin'. Shot one husband arreddy."

The dispatcher's wheezy voice came over the company radio then. "Got us a little bitty Birdwell here. Soon's you load her up, you can head on out."

"Thass a roger. We're a-hookin' her on up now," our captain sang-song back. The rest of us groaned up the stairs and out onto deck. Just this one more run and then some sleep. If we're lucky.

We'd returned just an hour before from that same rig, having made the run out in fog thicker than kapok, then coming back in seas rising up before one devil of a west wind. Men on the fuel dock scratching their heads wondering west wind? Who ever heard of a high west wind this time of year? Hurricane season, sure, you get some bad west winds then. But now?

The dock's crane loaded on the "little bitty" Birdwell, forty-one thousand pounds of high-tech radium camera. Theoretically too great a load for our little *Empress*. But hell, it was a short run, not more than ten hours out and back, even counting for delays at the rig. We could handle it.

Two passengers were aboard, the Birdwell operators, leaning over the radar for a chaw with the captain as I let loose our lines and we gunned out of dock.

Captain Corey let me pilot, my first wheel time on the *Empress*, while he curled up with Big Bettie in the passenger lounge. I watched the sunset glowing red and purple into the wheelhouse, colors melting on our windshield where the sea spray spattered. Even coughing and belching as she was, the *Empress* was a fast boat, twice the speed of the *Pride*, almost as fast as the *Buck*. She was trickier to handle than the *Pride*, too. There's no cutting through waves on a light boat. Best to get up speed and plane on the water, a new sensation for me at the wheel. I'd never been allowed to drive the precious *Buck*.

The waves that afternoon were peaking high, boiling up the channel from the jetties. The *Empress* fluttered and bobbed. I thought my hand on the wheel was responsible, but when Corey took the controls for a minute, he did no better, so he handed them back to me.

An armada of shrimp boats passed us on its way to shelter at our dock, more than a hundred of them. The wind was picking up, threatening to sail the small boats wobbling into our path as we came through Cat Island Pass.

When it was time for us to take our dog-leg turn east, through The Pillars, a pair of treacherous pilings that rise out of the channel like the tusks of a ruined mammoth, the wind caught us just so and whipped us completely around against our rudder. That's when Captain Corey should have headed us right back to the dock. Instead, he nuzzled Big Bettie and waved me on, out into the Gulf.

A wild ride. Climbing a tall crest, teetering there a moment, then sliding, then scrawing down a steep trough, burying our bow in green water. I jammed my feet against the dash, wide apart, to keep my grip on the wheel, to keep my seat in the pilot's chair. *WhooeeeeEE!* Going *uuuup!* Coming *dowwwwn! Ooo-eee-OOO!*

I cast a look behind me, down into the lounge. Corey and Big Bettie were strapped into the benches, and while she tensed herself, bug-eyed, for each wave smack, he nuzzled and groped at her vastness. I couldn't believe he was taking this weather so calmly.

We had not then left the shoals. When we did, we began to ride the real seas, fourteen-footers, the steepest I'd ever seen. When the *Empress* buried her nose, right up to the windshield, in the seas and a trail of seaweed stuck in our window wiper, I persuaded Corey to take the wheel. I helped Big Bettie get below where, I presumed, she would turn green in solitude. I furnished her with a plastic bag just in case.

The high seas played a terror game with our small unstable craft. Lifting her up on one roller, dropping her twenty feet head first into a deep trough, washing up over her bow, smacking her stern on their way out. Some of the waves could be counted on to catch her on her beam. Those were the worst. Cracks like Judgment Day when the *Empress* flew up out of the water and smacked hard coming down. Between the cracks, there were creaks from the hull under our feet. Slow, long, spooky creaks, Dracula's coffin creaks. Aluminum hulls aren't supposed to creak at all.

Most ominous, we appeared to be the only boat left on the water. The last shrimpers on the Gulf whipped past us floundering between massive waves. I was, for a moment, Dorothy in midtornado, peering out to see all Kansas whirling by.

And then the storm hit us, hit us hard with rain as dense as a waterfall. I'd seen some tropical rains before, but nothing like this. Maybe we shouldn't be going out in this storm at all? But then, I was new to the Gulf. I'd never heard of a boat turning around and heading for dock just because of a storm.

Even when we heard, on our last functioning radio, that our rig had shut down drilling for the duration, and the captain still kept driving, I did not at first question him. He was the captain. Like Guste, like Willy and Percy, he no doubt knew what he was doing. I did, after a while, ask him why we were running to a rig that was shut down. Once shut down, they'd refuse to offload us. I knew that much. So shouldn't we turn back?

"I never turn back," the captain declared. "I never did, and I never will."

The night before, the captain had told me he'd never creased another boat in all his years on the Gulf. A few minutes later we rammed the stern of a supply boat that was stalled in the foggy channel. Nothing serious, but Captain

Corey had lost face. Now he would make it up, man against the storm.

I could not be certain of his motives. I couldn't even be sure, bennies or no, that Corey was certifiably crazy; he was an old hand out here, wasn't he? I took a lighter tack.

"Well, you only die once, right, Cap'n? Might as well enjoy it." I lit a joint.

"Thass right. Might's well go out on a big one."

The conversation I had begun in hopes of turning him back we carried on to release our giddiness. Doom can be fun! Great comic paragraphs rolled off my tongue. But it wasn't the dope; the joint had died out after one toke. I was stoned on fear.

An hour later, and no closer to our rig, I'd fashioned a seat belt for myself out of manila rope and a screwdriver handle. Wrapped around my chair in the wheelhouse like a pretzel, I was as zany within as I was calm without. Whenever our windshield cleared of wind-driven water long enough for me to get a glimpse of the seas, I felt my eyebrows hit my hairline. The spring before, I'd seen the Potomac River at full flood, spitting up chicken houses and convertibles and swallowing them again. But here each wave was an angry river of its own, as powerful, as fearsome. Nature jumping the track. So this is it, I thought. So this is it.

Suddenly the captain jerked in his seat and screeched, "See that tree? Thass the biggest goddam tree I ever seen in my life, excuse my language."

"But, Captain, we're miles from shore. There aren't any trees out here."

"Oh, my dear, you're perfectly right. Whew! I must be *seeing things.* You know how those little yellow mollies do you. Now don't you worry. I ain't crazy—yet! Haha! I been in wuss than this storm, believe you me."

Worse than this storm, worse than this storm. I was strangely reassured to hear there was such a thing.

Again Captain Corey nearly nodded off, then jerked awake, scrawing the throttles to full forward starboard, full reverse port. "Watch out!" he screamed. "We're gonna hit the ditch!!"

Ditch?

"Well, whew! Guess I was hallucinatin' again, heh heh."

Avoiding that ditch, he'd thrown us into a dangerous side sea. We got swamped twice before he could pull her around again. I heard our bilge pumps kick on. Sam, our engineer, must be awake.

Sam and Mighty Mouse joined us in the wheelhouse then, hauling themselves up the ladder from their quarters. Both were pale with fear, Mighty Mouse bloody nosed. He'd been sleeping, after too many hours awake, when the *Empress* hit that first killer wave and blasted all two hundred and eighty pounds of him up to break his nose on the ceiling of his bunk. Too tired to notice, he drifted back to sleep until Sam roused him.

Sam, equally exhausted, peered through our windshield and saw a giant roller heading for our port side. "What's happening!" he screamed. "What are we doing here?"

"We're too near the rig to turn back now," the captain answered. "We'll anchor and tie up there, ride out the storm."

Sam howled. "The anchor winch is busted and you know that very well yourself!"

"Don't worry about it," Captain Corey said, as the *Empress* bucked and spun on her stem. "We'll rig something up. Right now why don't you and The Mouse go down to the engine room and see if the bilge pumps are doin' us any good."

The two men made their way astern then on all fours for safety. Even so, I heard them grunt and scramble with each sock of wave on hull.

Captain ordered me down to the galley to fetch him some coffee. On deckhand reflex I untied myself and started out of my seat. Too late I realized no coffee could be made in seas of that magnitude. Too late: the *Empress* shimmied down the long side of a trough and I was grazing the overhead with the back of my neck, then sprawling over the radar with my chin on the captain's shoulder. Thrown next against the port-side door, I grappled myself back into my chair.

The stink of fear was all over me, rancid, sharp, personal.

The poor Birdwell men, forgotten back in the lounge, were vomiting where they sat strapped into their seats. Big Bettie's moans seeped up to us, rising occasionally to a reedy crescendo. I noticed that Corey had traded his usual rubbery clown face for a stiff bennie grimace. He was fighting the steering, and it wouldn't answer. No wonder he looked so ghastly. Losing steering is halfway to being set adrift. Adrift, taking the seas on our beam, we might not stay afloat for half an hour.

"Lucy, you and Sam better go top up the steering fluid," the captain said.

In heavy seas it is just barely possible to steer one of these light boats by throttle alone. In this case, the rudders were working against us, either jammed in one spot or flipping wildly. So Captain Corey grappled with the wheel while Sam, just returned from the engine room, roped himself to me. The two of us set off for the back deck to check on the steering fluid. We had no life jackets, at least none with functioning buckles. So we would work like mountain climbers, tying, untying, and retying ourselves to the deck rail. That was the plan. Eighteen- to twenty-foot seas now, ninety-mile-an-hour winds gusting up to one hundred and ten, the rig radio had warned on its last broadcast. All Sam had to say about it was, "It's never been this bad. It's never been this bad."

We were crazy to go out on that deck. But hell, we had to try. What else was there?

Out the door the first sting of windswept seawater drove straight through my wool peacoat, through my sweatshirt, through my T-shirt to my flinching skin. Just. Like. That. A paper towel I'd stashed in my inside-most pocket (I thought I'd use it to wipe the rain out of my eyes, I was that unprepared for the reality of this storm) was soaked through in the first moment. The power of that storm. The sheer awesome gut-wrenching power of wind and water gone mad.

We crawled on our knees, hitching ourselves along on the clumsy lines that bound us to the rail, losing our purchase on horizontal with every wave that broke over us. Gasping up from under one wall of wave, we lifted the hatch cover off the steering fluid fill and saw that the fluid was already full. Then the next wave flattened us to the deck and tore at us with fearsome power, sucking us into the bulwarks. Let me put this another way: Have you ever played on a seashore when the water was rough, fighting the surf just for fun? Felt the suction of that surf lift you up, plunge you under? That's the feeling. Only we were not comfortably close to shore but thirty, maybe forty miles out, with just this ruined wreck of a boat to cling to.

The wind yowled like a chord organ, wavered like a concertina, whistled, screamed, roared, wept. If I'd ever known it at all, I'd forgotten that storms can moan and cry like that. I looked up, between waves, and saw that our radio antenna, a thick fiberglass whip, was blowing over almost flush with the deck. That wind. That killer wind.

Since the steering problem lay not with the fluid, then it must be the rudders. That was bad news. Rudders are vital organs. Timing it so we'd lift the engine room hatch between waves, we lowered ourselves into the power plant room from deck and grappled our way over its sharp edges to the rudder compartment. The bilge we floundered in was up over our knees, and the waves in the bilge very nearly had whitecaps on them. Scary things to see, nothing wherever I looked but scary things to see: the familiar gone strange, the safety of the boat become a powerful menace. The thrill was gone.

Our rudders, we saw, were jammed against opposite walls, their pins lost somewhere in the heaving bilge.

We slipped and slid, grappling with cold hands under greasy water until we found the pins, wrenched the rudders back, and stuck the pins in place again, hammering them with the raw heels of our hands. Sam, working against the antigravity of the bucking boat, hammered hard, saying squeaky and incoherent prayers not quite under his breath. I understood him perfectly.

When we reported back to Captain Corey, he had the steering back under control again but his drug mania seemed to be peaking. "Either I get a cuppa coffee *now* or *you* [indicating me] get offa this boat." From the time I first decided to run away to sea, I'd hoped someday to be involved in a mutiny, on the Fletcher Christian side. This was the first provocation I'd suffered, and, alas, just the wrong moment for it. I had enough to worry about, just trying to cling to the doorway and stay on my feet. Not one of us was confident that any but this whacked-out drugged-up *experienced captain* could run ahead of those evil seas back to our dock. I kept my silence and the captain's threat died.

We did make it to the rig, but as I feared, the rig wanted no part of us. The Birdwell men groaned in terror when they heard that the crane operator up on the rig didn't want to risk lowering the personnel basket to lift them off the boat, let alone lift off the giant Birdwell box that was contributing so nicely to our unseaworthiness. Captain Corey carried on a wild, unreasoning argument with the drilling company's tool pusher over the radio. At last the basket came, but only for the Birdwell men. The box stayed aboard.

No sooner were the Birdwell men plucked off our deck than our steering went out again. "What are we gonna *do?*" Sam cried, his face stretched into a demon mask, the scariest thing I'd seen all night.

Our little *Empress* hung on the seas, in peril and without power, just to the weather side of the rig's enormous saw-teeth, each tooth as tall as a man. And then she plunged under the rig.

The captain broke down then, cowering on the sloshy wheelhouse floor under the dash as the shadow of the rig passed over us. It is theoretically impossible for a small boat to survive a bashing on a rig. It'll hang on the sawteeth and open up like butter to a hot knife. It is, theoretically, even less possible for a boat to pass under a rig without getting bashed up. I'd heard about the one that tried. Mighty Mouse and I pulled our heads into our shoulders, as if we could shrink the boat. Maybe that's what saved us. Because the *Empress* bobbed up, miraculously unharmed, on the other side of the rig. Sam scuttled down to the engine room to check out the rudders again. Mighty Mouse and I led the captain to the stern controls above our back deck, the only wheel with good visibility. We hoped he could steer us well away from the rig using the boat's throttles to alter our course.

So, the back deck again. And this time, against the power of the wind and the waves that pushed and pulled me, I wrapped my arms and legs around a vertical stanchion. Still, I swung with every wave, nearly lost my grip with every rush of the water's fury. I kept my head down when I could, to minimize the rush of salt water up my nose. I tried to think of something I could promise God I'd give up if He'd get me out of there alive. Through the bone-chilling howl of wind and the wet belch of our futile engines, I could hear the captain screaming for the same favor. Captainly macho, like our ship's flag, had been stripped away by the storm.

The rig's standby boat, at anchor not far away, suggested by radio that we come alongside and tie up to her. That would be fine, if we could find her. The voice of that unknown skipper sounded so much like Guste's that I jerked for a moment to full attention, then felt my despair fully. No, I was not safe and in the care of a sound captain. In fact I was liable to die, for real, in some gruesome manner, right there, right then. Unless we could find that boat, that *Miss Crickie.*

Sam reported back. The rudders were pinned again, but he didn't know for how long. We'd better hurry and find that boat.

The captain insisted that he must sleep first, sleep or die. But Mighty Mouse surrounded him, braced him, murmuring God knows what into our captain's ear, protecting our leader from the wind and the water so that he might attempt to zig and zag us the mile or so to *Miss Crickie.* And we did make it that far.

Stern to stern with the standby boat, one more miracle was called for: that someone volunteer to go to our stern and hurl a heavy line across to the other boat. A simple enough maneuver on a sunny day, but now! Every roller lifted *Miss Crickie*'s stern twenty feet above our own. (Look at the size of those propellers! I'd no idea boat props were so big!) Then, as she dropped into the trough behind a wave, we'd rise as far above her.

Mighty Mouse wrapped his coat around our captain and staggered bare-chested to the stern with our best line. I sucked in my breath and held it while a wave rolled over first him, then me. I looked up expecting to find him gone, sucked away into the madness of the storm. But he was standing firm, hurrying to snag the *Miss Crickie* before another roller swamped us. After a couple of near misses, he caught the bitt in a perfect loop and tied us fast. I made myself believe it, because it was true: we were safe. We would not drown tonight.

Mouse said he'd take the first line watch, and the rest of us began stripping off layers of clothing, clothing that was rigid with ice. Funny, I hadn't even noticed the ice before.

But by the time we'd all gathered up our sleeping bags for a short nap in the passenger lounge, Mighty Mouse was hollering, loud. A monster wave had wrenched the rusty bitt off our back deck and set us loose from *Miss Crickie* again.

What followed was a more practiced repeat of the earlier tie-up, with *Miss Crickie* sending her deckhands out to lasso our portside bitt with two hefty lines. Again we set up watches, mine first this time, and I relaxed. Now we were really safe.

Ten minutes went by and then the lines parted. Our *Empress* was adrift in side seas, badly awash.

When I shook the captain awake, he went for my throat before he regained consciousness. Once awake, he ordered me up onto the wheelhouse roof to aim our spotlight, so that we might find *Miss Crickie* again. "On your way down," he said, "see to the lifeboats."

On my way up the ladder to the wheelhouse roof, I collided with a dead sea gull washed square into my face by the crest of a wave over the wheelhouse. But I had no time for omens, not then, not when life could not be trusted to last another hour. Fight or die, that's all there was.

I wrestled with the spot, lashed it in place, then considered what might be meant by "see to the lifeboats." Then I did see. Someone had stowed loose gear and tangled lines in the rafts. I wrenched that mass of junk out of there, let it fly with the wind, and was chilled to see how fast it disappeared into the storm. Gone, just gone, clearing the lifeboats so I could see how the sun had long ago melted the stacks of boats into one single orange polyurethane plosh.

Not that it mattered. Keeping a perch on one of those flat little rafts was beyond even imagining on seas that crested twenty feet above our decks and swallowed up coils of rope in a split second.

On our third attempt to tie fast to the other boat, the inexorable wind won out. The lights of *Miss Crickie* disappeared behind a curtain of rain as we were driven into the blackness of night. We would have to try a run to shore.

And so we ran, doing our best to stay steady and upright atop one long wave at a time. Lose one wave, climb another before the next one roared over our stern to swamp us. We made surprisingly good time. Pretty soon we were congratulating ourselves for surviving. As we neared shore, I radioed for help. No answer, not even from our rig. Eerie, eerie feeling when it seems, if only for a moment, that you are the last of your kind in all the universe. I radioed for help again. No answer. What few boats remained on the Gulf were in as much trouble as we, I guessed. Still, I wished somebody knew where we were. My mother, for instance. The currents in the shoals twisted us, leaf in a tempest.

No sooner had we sighted the dim green light of our number one channel buoy than the current caught us and ran us aground. All of us shouted, cursed, enraged that this night was not over yet. We were still miles from safety. The *Empress* rocked savagely with the massive waves that battered her, driving her hull deeper into the suction of Louisiana mud. Captain Corey ground the throttles forward, then reverse, then forward, hoping to rock us out like a car stuck in a snowbank. The rest of us stood as upright as we could, our faces bathed in the green light of the radar.

"Do something!" Corey shrieked. Damned if I could imagine what.

Arrived then our first piece of measurably good fortune that night: a battered old utility boat on its way out (*Out?* Are you crazy, man? Do you know what it's *like* out there tonight?) paused long enough to shoot us a line and pull us off the mud. But surely they didn't mean to go offshore!

They shouted to us that they'd heard a distress call from a pumping platform, four mechanics left on it with no shelter; they had to go try to get those men off. The storm was dying anyhow, so what the hell . . .

Off the bank and running more carefully for our light, we slapped one another's backs and hazarded the first sighs of relief. Then we smelled smoke. From the engine room.

Our exhaust ports had clogged with mud when we ran aground. Now that we were under power again, the rubber ports were going up in smoke from our superheated, bottled-up exhaust. Billows of burned-rubber smoke puffed up into the wheelhouse. Big Bettie, long unheard from, bolted up the stairs, choking. First one main engine, then the other, coughed and died. Captain Corey sagged into Big Bettie's arms. He was through trying. This last blow was too much.

Sam ran ahead of us to the engine room while Mighty Mouse and I gathered up every fire extinguisher abovedecks and lugged them all below. We found Sam standing well back from the smoke, shooting foam in the direction he presumed the fire to lie.

My boat was burning! And Sam was letting a little smoke keep him from rushing to the source of the blaze to smother it out. "You gotta go in there," I shouted to Sam.

"Too smoky!" he shouted back.

I could sense that the fire had bounded forward into the engine room. I could almost see the flames . . . yeah! There they were! "We've gotta do something," I cried again.

Sam leapt forward with one extinguisher, and I grabbed up another one, thinking to hand it to him when his ran out. I could hear a choking behind me; Mighty Mouse. Was I waiting for him to take in a third extinguisher? He was in no shape to try it.

So I went. I duck-walked into the smoke and stayed there, squinting my eyes shut, not breathing either, sending a *fwoosh* of chemical foam at what I hoped was the base of the flames. When the sting of smoke crept in through my tear ducts, I inhaled on reflex, then coughed and fled to smoke-free ground. I guess I was no hero.

I did go back in again, but it was Sam who stayed the course, dashing in with the foam and out to breathe until the fire was out. Still, my boat had been afire and I'd responded. I did not, at least, Blow It. And what was that about *my* boat?

Not my boat at all! This rotting hulk, this verminous scow! This leaky, stinking, stove-in, good-for-nothing-but-scrap-at-the-pound ruin! I mean, shit in a bag!

And she'd made it. Shame that she was, she'd brought us through alive. For a moment I dreamed that I would chip and paint her pitiful hull, scrub and scrape her every miserable inch to shipyard newness. I'd make her want to live again . . .

A brave little tug interrupted my fantasies, pulled us off the bar at Cat Island, towed us into dock again. The storm was blowing itself out; the rain had passed. The four of us rode into dock in the wheelhouse, curiously silent, watching dawn lift the waterfront of boats out of shadow. We looked on life in Dulac, Louisiana, with newly benevolent eyes. Or I did, anyway.

And then we all fell out, dead to the world for a full day and a night. I must have mulled over my experience while I slept, because when I woke up I had a new framework for it, not a romantic framework at all:

I'd earned thirty-nine dollars risking my life to deliver two Birdwell operators to a dirty old offshore oil rig. If I'd lost my life in the attempt, my kids would have made ten thousand dollars. Ten thousand dollars would hardly cover the price of a Pontiac Grand Prix these days. My life is worth more than that to me, I realized.

The Coast Guard would be coming aboard to check out the *Empress* now. She'd wind up in drydock for sure, if not in the scrap yard. I packed my bags and kissed my crew goodbye. They were packing, too.

Big Bettie, who staggered off the *Empress* when I did, took me aside. "What for you got to be out dair on dem boats, *chère?*"

"I don't have to be out there, Bettie. I like boats, that's all."

"Den, *chère,* you crazier dan de men!"

Part Three

LITTLE HORRORS OF LOUISIANA HOME AND HIGHWAY

15

Atoms of gray rain drifted out of a smudge of winter sunrise while I stood over my gear at the Dulac dock taking shaky inventory of what was mine. I was alive, my singes and bruises would heal soon enough, I had a pocketful of money and more seabags than I could carry. By any other sailor's timetable, this was the hour to head home.

Home. I'd never feel at home on land again. Even the *Empress's* narrow escape from the storm couldn't shake my dread of land. I belonged on the boats, or I would, someday, belong on some boat. What soured my mood that morning was the possibility that I'd be roaming from captain to captain, from boat to boat, maybe for another month, maybe even for years, before I found the boat where I belonged. I'd run through five captains already: one incompetent, one psychotic, one just plain unfriendly, one too close for comfort, only one Goose. Finding another captain with Guste's good qualities was turning out to be harder than I'd thought. I was, that morning, too tired to start my search again. I slumped back down on my bags and sat there a while feeling blue and empty.

A three-quarter-ton pickup pulled up next to me and its driver offered me a lift to points west. There was an idea: I'd check in with Guste, Cupp, the *Pride*. The trucker said he'd take me there if I didn't mind stopping over with him at his trailer in Thibodaux. We could sort of rest up, he said.

Yes to the ride, no to the rest stop. If this man so much as touched me, what with the mood I was in, I'd probably rip his heart out with my bare hands. I told him something like that. With those terms and conditions agreed upon, we headed out.

The bumper sticker on the truck's rear window read: Oilfield Trash and Proud of It. The driver's name was Bill, he told me, but I could call him by his CB handle, Cherry Picker. Cherry Picker was hooked on his CB. "I allys got my ears on," he said.

The radio fizzed: "I'm a-lookin' at a red pick-'em-up truck with a A-frame on it there. That you, Cherry Picker?"

"Thass a roger. Who be talkin' at me?"

"This be Quack-Quack talkin' at you. See you got yourself a sweet thang."

"Got a beaver, thass a roger. Seen her hitcherhikin', she be goin' for the Intercoastal."

I interrupted. "I am not a beaver. I am a woman."

"Don't be gittin' your purty little ass in a uproar. Beaver's just a word we call women on the CB, like we call them state troopers bears."

"And what's the CB slang for men?"

"Buffaloes."

"No, no, no. That's not right. If you're going to call women beavers, you have to call men roosters. Fair is fair."

"Yew don't unnerstand," Cherry Picker whined. "Women *is* beavers. Men *is* buffaloes. If I was to call a man a rooster on the CB, they wouldn't know what I was talkin' about."

Deep in my foul mood, I clamped my arms across my chest and peered out the window. A flock of long-legged white egrets, damp with morning rain, picked through a roadside dump, stepping over rusted beer cans and plastic milk cartons. Plump nutria darted from mound to mound in the shallow oil-slicked marsh water. Oilfield trash and unwary of it. Where was this running away to sea getting me?

Then came a loud metallic *bank!* Cherry Picker had

thumped the ceiling of the truck's cab with his fist. "Goin' to sleep?" He grinned, evilly.

"No, just looking out the window. Just thinking." I resumed my blue reveries. Brahma bulls stood beyond barbed-wire fences, stolid, some of them with heads up to meet the bare mist of rain. An endless soggy plain stretched out brown beyond them to the horizon. I even envied those cattle. They had their fences, their limits. They wouldn't get lost or entangled in messy adventures. Where was I headed?

Another startling *bank!* I jumped in my seat.

"You think too much, thass your trouble," Cherry Picker accused me.

I glared at him.

"Well, pardon my ass," he snapped, and lapsed into a sulk.

I would have liked to treat him to my riff on the rites of courtship among Dixie aborigines like himself, how a male of the species announces his romantic intentions by throwing a good scare into his chosen female. But Cherry Picker wouldn't have understood.

Musings of that sort broke my thread of self-pity; my homeless blues passed off and in the icy privacy his pouting gave me I collected a spiteful list of misguided male vanities from passing bumper stickers: "Divers Do It Deeper," "Welders Can Do It In Any Position," "Helicopter Pilots Keep It Up Longer," "Riggers Do It With A Big One," "Schlitz Lovers Do It With Gusto," "Coonass Do It Best 'Cause They'll Eat Anything."

Good, I thought. Let them do it and do it and do it, they could do it to death so long as they weren't doing it to me. I was on my way to see my friends.

I found Guste in the galley of the *Pride* boiling up a chicken for his lunch. "Got me no mo' gallbladd', *ma fille*. 'Bout time mah Yankee cook come back."

We traded recent sea stories—I had an adventure of my own to tell at last—and then Guste said he'd lost his pretty

Florida cook to a local crew boat captain. If I wanted my old job back, I could have it.

Foundling though I was, I don't think I felt even the slightest tug of temptation; I was a deckhand now, an experienced deckhand, a good one.

Over lunch, Guste told me he'd been offered a captain's job aboard a deluxe new supply boat being built for Petrolco. Petrolco told him he could hire his own crew and name his salary, but he'd turned them down. He was too old to change boats now, he said. But if I thought I'd like to come along with him, maybe he'd do it after all.

By the end of our lunch, he'd made up his mind to go with Petrolco, pick me for a deckhand, train me up for mate. He'd see if he could talk Cupp and Fred Fatigué into coming along. I might have to wait a week or two while he worked it out; could I do that?

Just that suddenly, all my troubles were over. In a matter of weeks I'd be back with my captain, my friends, at sea. Sure I could wait.

That night I slept on the *Pride*'s galley bench. Next day I hitched a ride with Cupp's crew change to my cabin by the swamp. Guste told me to stay close to the phone. He'd be calling me soon.

I heard a pop and the snake froze as fast as that, one moment cutting an S of flight through winter-browned grass, the next dead. Shot behind the ear.

"Got'm," Cupp snapped, satisfied.

A snake is hard to kill, Cupp said. "Got to nail him behind the eye first off. Blow'm in two and the brain half'll run off on you before you kin line up your next shot."

I cringed. Cupp handed me his .22 rifle. "Here. You try."

When I was five years old my family lived on the bank of the White River, at a place called Trail's End, in central Indiana. Trail's End was alive with snakes. My mother taught me not to fear them.

Once when my father was away for a few days, my mother noticed something tugging on a fishing line he'd left in the water. When she pulled up the line she found a water moccasin on it. The fish hook had speared him through the jaw. I watched while my brave mother grabbed the thrashing black snake, removed the hook from his poisonous jaw, and threw him back into the water. Me kill a snake?

I showed Cupp my field guide to reptiles, compared the snakes he'd killed with their portraits. They were all harmless species: king snakes, garter snakes, black racers. Okay, so one was a cottonmouth, but it was just a *baby.* Hadn't Cupp ever heard of live and let live?

Cupp snorted. "One-a them snakes take a chunk outa your laig, what you gonna do? Read to 'im outa your book?"

Cupp had his ways, I had mine. I gave up following him on his hunting trips and pulled back into my house to wait for Guste's call. While I was waiting, spring came. It was the rain that did it, warm rain tasting alkaline. It rained for one whole week, night and day, first pounding, then leaking. The hot sun came after. Spring rose like a wall. As if they'd been boiling just below ground, wild grape vines twisted up barren trees and slapped a primary layer of green over the swamp behind my house. A stingy little stick of a tree by my driveway blossomed into a pink mimosa. Within an hour a literal hundred hummingbirds appeared, drinking from its gold-fringed blooms. I'd seen exactly four hummingbirds in my life. I sat under the mimosa, watching the hummingbird extravaganza, letting the sun bake my bones. A muddy rock at the edge of the bayou baked with me. After a while it lifted itself on four legs, stretched out its neck, and hot-footed it for high ground. It was a snapping turtle that lunged at me with hoarse snarls and clapping jaws when I followed it with my own snapping camera.

At dusk, moonflowers opened. White herons drifted to earth under the newly green willows that grew over the ditches. Armadillos snuffled around my back door. I could barely register one wonder before another appeared: raccoons rattling my trash cans, giant fruit bats swooping under my neighbor's backyard spotlight. Sheep frogs started up their droning *baa-baaaaa*. As Cajun folklore had it, sheep frogs cried for the rain. And always the rain came, driving out of the Gulf fifty miles away, drumming on my tin roof.

When the rain rattled off again at morning, it left behind another humid rush of growth. Fern croziers broke black earth. Ribbon grass sprouted fat striped shoots. The spring crop of tree frogs clung to the screen of my front porch, each one of them no bigger than a thumbnail, the mass of them so dense they blocked out the daylight. This scene had a horror-movie look: invasion of the frog-people. But when I came closer, I saw their thousand eyes all deep and black and certain, their throats beating out a single pulse. The pulse was an urgent message, and each throat seemed as sacred as my own.

This bayou world was a steamy hothouse of life. Of death, too, of course of death, too, death as relentless as this rush of spring. Life I could live with. Death I would have warded off. I daydreamed of enclosing my half acre with a giant netting cage, posting it off limits to hunters like Cupp, to predators like the sparrow hawks who raided the warblers' nests in my honeysuckle. When I saw the first warbler fall, I counterattacked with a broom. I shagged a rock at a lone bobcat who skirted my yard with a bleeding nutria in his mouth. When a flock of egrets came to harvest the sheep frogs, I waved them off, pounding two skillets together. One egret slipped past me. I watched a pair of green legs squirm between his yellow blades of beak.

I knew I was on shaky ground, defending one wild creature from another. But where man the hunter was concerned, I felt sure I had it right. Old Man Duson, my neighbor across the road, poacher king of Lafourche Parish, offered me a stiff sackful of baby egrets from his freezer. "Makes the richest stock for gumbo," his wife said.

Our entire neighborhood, four-generation coonass families and oilfield transients like the Cupps, turned out in the after-dinner hours for snake shooting and crawfish grabbing. Men and boys leveled their gun barrels, squinted, shot, then kicked the corpses around, playfully. Women and girls made it their business to nab every crawfish that showed itself above ground. The crawfish, called mudbugs here, arched their backs and stood their ground, squaring off against the Cajun women who stooped low over the ditches with crawfish pots in one hand, fingers quick as death in the other.

Late into the night, long after the men had gathered up their trophies and the women had boiled their mudbugs alive, I lay awake, twisting between sweaty sheets, hearing migrating crawfish crackle under the wheels of trucks on the highway outside my window. On the worst of those nights I imagined I could hear bobcats plundering the nests of infant rabbits, crows dipping their beaks in mockingbird eggs.

I'd always bought hamburger in neat plastic packages. Now it was all over the blacktop, a road I would otherwise have used to walk off my insomnia. Under the cane farmers' mercury lamps lay split armadillos, broken 'possums and 'coons, cracked and leaking turtles, stiff black dogs, flat orange cats.

I woke one morning feeling blissfully light, as if I'd somehow passed the crest of the horror. I would face it, accept it for what it was: the way of the world. Then I stepped outside to my mailbox and met a new *danse macabre* at road's edge: the freshly concussed carcasses of young barred owls, wings still beating out flight in the gusts of wind from passing cars. I retreated indoors again and pulled all the curtains shut. But not even my four walls could hold back the horror. A mouse drowned in my dishwater. A tree frog leapt into the bedroom doorway as I passed. He was flattened underfoot—my foot, my innocent foot—before I could shift my balance to save him.

In all innocence I slammed a window shut against a thunderstorm, then saw, too late, an inch of electric blue newt's tail twitch once and be still.

I befriended a pair of chameleons that came to lick up

midges from my screen door. They stayed longer every day, drawn by the music of my record player. I was as innocent as Snow White, friend to all helpless creatures: I knew the chameleons somehow sensed this. I lured them onto the backs of my hands and the three of us spent long midday hours together, tranced up by the music, safe from the storm of death and dying that raged just beyond my walls. I'd been alone and outcast for many months. Now I had an audience for my best stories, deserving recipients for my Snow White love. I gave them names, Warren & Wife. My mistake.

A week into our friendship I found Warren flattened dead just outside my front door, the door I'd let spring shut behind me a minute before. I made tearful apologies to Wife, but she didn't come to see me anymore.

Out in the yard another day, I lifted a paint can lid and found a poisonous coral snake—or was it his look-alike nonpoisonous cousin?—beneath it. As I ran for an axe and my field guide to the reptiles, a fat orange-headed salamander ducked back into his hole in the siding of my house. He knew, he knew; I was man the hunter, no better, no worse.

That fact didn't come clear to me until I met a particular dragonfly, just one of dozens who tickled me when I lay sunbathing in the yard. This particular fellow, whom I recognized by a foggy, asymmetrical pattern on his back, turned up to inspect my work when I painted the house. His head ticked around curiously, as if he were bemused by my sweaty earthling pursuits.

I made him my hero when he snatched up the horseflies that came around to suck my blood. Here was some natural carnage I could approve: the way his neatly retractable forearms positioned the horsefly's squirming body for the first firm clap of mandibles. My dragonfly seemed to nod between bites, thoroughly satisfied with The Establishment. He was just a baby, needle thin when I met him. I watched him grow fat and strong.

He didn't live long. I found him caught and sucked dry in the web a banana spider had woven overnight between my back porch pillars. In a fit of rage I soaked the spider with poison from an aerosol can. I called him murderer.

And then I saw the true nature of my innocence. Who was I to take sides with God against God? God is food. God is hungry. God, or at least Life, is consuming Himself at every living dying moment, and my half acre of Sixty Arpents Road was just one tasty crumb on His worldwide dinner table.

Gradually I learned not to flinch when He came as a heron to pluck off His chameleons in the grass. Or when, as an owl, He tasted His toadlets. Not even when His hawk's fine talons pierced His mother mouse, whose pink darlings nested in the stuffing of my sofa. She must have shared His appetite when she ventured out. God is hungry, God eats. Amen.

Once I settled that with myself, I could watch with a new benevolence when God, got up as a water moccasin, draped Himself across the center ridge of my flooded driveway to take communion with the crawfish streaming by. I even tingled with appreciation when His kingfishers hurled themselves into the bayou like knives and came up smacking His lips.

I remembered the boats, how my heart jumped up in my throat when I saw a pair of does standing knee deep in the Freshwater Bayou, ignoring an alligator who yawned at them from his mudhole in the bank. When I pointed them out to Guste, he'd said, "Yah, yah, plenny dem down dair. De woods, she fulla meat."

Even when I'd resolved my god-eats-god dilemma, I didn't rest well. Nights, under the drumming of the rain, I wandered through my customary dreamscapes, and no matter where I found myself, in the twisty corridors of a ramshackle mansion or the echoing emptiness of a high school auditorium, I'd remember, suddenly, that I was missing my boat. The boat would appear then, always just over my shoulder, leaving without me.

I'd whirl, stopped cold for a dream instant watching the

boat pull away from dock with its propwash churning the dark waters between us. Then, galvanized by a jolt of panic, I'd leap, right hand and right foot outstretched, straining for the safety of the boat. With the leap came a hushed, elongated moment, a bare whistling in my ears. I'd closed my dream eyes tight, could not know if I'd attain the boat, the safety of the boat, or fall into the murderous wheelwash.

Then I'd bolt awake in the noisy Louisiana night with my heart hammering in my ears, drowning out even the drumming of the rain.

One morning, after a month of dreams and dream leaps, of waking stone frozen with fear, of waiting for Guste to call, of packing and repacking my fast-mildewing seabags so I'd be ready on a moment's notice, I locked the house, carried my bags to the end of the driveway, stuck out my thumb, and hitched a series of rides to the *Pride*.

The Goose, when I found him, seemed embarrassed, defensive. It wasn't that easy, he told me, for him to quit the Watercraft Company. He had a nice bonus coming at the end of summer, his wife was kicking up a fuss about his pension plan, and the offer from Petrolco didn't look as good as it had. I was still welcome to the cook's job on the *Pride*, but . . .

I didn't try to hold my friend to a promise he regretted. But I knew now that I couldn't live on land again or in a galley either. Time to look for another job.

Guste knew of one. Tied up alongside the *Pride* was the *Condor*, a supply boat as good as the *Pride*, Guste said. Just that morning the personnel man for Byrd Marine, owners of the *Condor*, had been sniffing around the dock for a crewman. Guste would take me aboard her to introduce and recommend me for the job.

Part Four

UNWELCOME ABOARD

16

The Condor *was* one of the rare boats that do not have a soul. I knew from the first I'd never even learn to like her. Her hull had an awkward profile, square in the beam, too high in the water. Worse, she was architecturally inhospitable, about as welcoming as a thruway rest-stop bathroom: empty, hollow, painfully white inside and out. Overheads, bulkheads, bulwarks, hull, all white, the stark chalk white of white shoe polish, of polystyrene coffee cups. Indoors, banks of bluey fluorescents accentuated the sensation of bonechill brought on by the overworked air conditioning. As Cupp would have said, she got me on my nerves.

What's more, I must begin as her cook. Only for a week, the Byrd Marine man promised. Then, thanks to Guste's recommendation, I'd be allowed a trial week as deckhand, the first woman deckhand with Byrd Marine, probably the first on any full-sized supply ship on the Gulf Coast.

Try out? This personnel man had been scavenging the docks all day for any piece of crud with a Z-card. I'm sure he would have dumped said crud aboard the *Condor* and felt confident the arrangement would work well enough. Here I had months of deck experience, good sense, a top reference, and way more gung ho than was good for me, and I had to beg for even this demeaning trial.

"Call it an experiment," the man said. Other women had applied to Byrd and been rejected out of hand. "I'm not ashamed to admit it," he said. "I'm prejudiced. But if you prove to me that a woman can pull her own weight out here, we'll take a look at the other gals who want to try."

Pull my weight. I wondered if I could. This *Condor* was no snappy little crew boat but a two-hundred-foot supply ship twice that size, with lines twice the weight and bulk I'd grown accustomed to. She'd be working a standby job, duty I knew nothing about, and most of the deck maneuvers would be new to me. She had two giant mud tanks to clean and an anchor chain to stack. Now that it was May, summer, the major work on deck would be chipping rust, grinding steel, painting the boat's entire exterior surface—endurance work. Maybe in my month at home I'd lost my new muscle and my knack for lines. Perhaps I'd fail.

Then again, could I afford to back off from an opportunity no woman had been given before?

"Don't worry about me," I told the company man. "I can handle it."

Wonder Woman was back.

Not for nothing do they call rice "coonass ice cream." When my new captain and I sat down together to map out the menu for my week as cook, he told me he expected to have rice at every meal. So it was white beans and rice, red beans and rice, stuffed roast and rusty gravy and rice, pork chops and rice, chicken and rice, gumbo and rice, sausage and rice. If I'd needed a reminder that I never wanted to work in a galley again, this would have been it.

At least I was back on a boat and sleeping well again. My nightmares simply vanished. And then the week was up. My Cajun captain and his crew went back to their regular boat and I threw my apron over the side: never again. I was a deckhand now, working aboard the biggest of boats in the oil fields, so a full-fledged top-of-the-line deckhand. A trial deck-hand, though.

The captain in charge of the *Condor* for my trial week arrived in the rig's personnel basket, swinging down out of the white-hot sky like some misbegotten Virgin of Guadalupe. He had about half a pound of chin on the end of his face, and he looked to be the very flower of white Western civilization: white skin well shaven, narrow white-blue eyes, a rich crop of whitening gray hair combed into a southern senator pompadour above a classic admiral-of-the-ocean-seas profile. This was Captain Billy Flowers, a retired Coast Guard captain, a perfect match in his dress whites for the inhospitable whiteness of the *Condor.* I noticed that his pearly dentures didn't fit him any too well.

He called the crew to attention. "Atten-*shun!*" I nearly giggled; this must be an oilfield first, a mud boat captain in dress whites reviewing his barefoot, red-eyed, raggedy-assed troops.

"Straighten your shoulders! Suck in those guts! Now, where's that woman thinks she wants to work deck?"

As if there were more than one woman aboard, I stepped two paces forward. Two could play his loony game.

Captain Billy inspected me for a few long moments. I knew because I'd seen a lot of war movies that I was to keep a straight face straight ahead.

"You want a fair chance?" he asked. "I'll give you a fair chance. You'll work right alongside the men, every bit as hard, every bit as long. And I won't cut you no slack, neither. If I hear you complaining and crying, I'll put you on the first chopper back to the bank. You got that?"

"Sounds good to me, sir."

"We'll see how good it sounds to you a week from now," he warned, winking at the rest of the crew.

"Fair enough, sir," I deadpanned.

He sent me redcapping his luggage to his stateroom and told me to unpack it for him. I'd never heard a captain ask a deckhand for such personal services. Still, I relished the glimpse of his private life this duty gave me: for wardrobe, Captain Billy packed jockey shorts, a garment for which I harbor a perhaps irrational loathing, and identical bubblegum-blue polyester jumpsuits patched with an American flag over his heart. The captain's personal cologne was, of course, Old Spice, his shaving implement a wicked-looking straight razor. For shipboard recreation he'd brought half a dozen well-thumbed Lance Horner studfarm slave novels.

I was thumbing through the slave novels myself, noticing he'd underlined certain steamy passages with a ball-point pen, when he marched into his stateroom, plunked himself down on his bunk, stretched out his right foot, and ordered me to remove his boots. "Remove my boots."

Remove his boots?

I begged off, lying that I had to make a quick trip to the head. No southern gentleman would question a lady's toilet needs.

Then Captain Billy called his deck crew, all two of us, to the galley for an introductory lecture. "I expect you to hit the deck at sunup and stay there until suppertime. There will be a thirty-minute lunch period at noon. Otherwise I don't want to see your face in my face or your butt on this bench. And while you are out on deck, you . . . will . . . work. I'm going to hear those paint chippers chipping twelve hours a day. Got that?"

Next Captain Billy reminded us that although we would refer to him as captain, sir, we should ever bear in mind his official merchant marine title: master of the vessel.

Yes, massuh.

"Dis-*missed!*"

I could see it coming: Hell Week.

Boat acoustics are peculiar, what with steel walls, narrow stairways, currents of Freon-blasted air. That night while I polished the wheelhouse brass I overheard Captain Billy, way down in the galley, telling the ship's mate that seeing as he, Billy, was one of the world's finer judges of character, he hadn't been fooled by "that woman," not a bit. Sure, she was trying to look sharp, and she'd put in a fair afternoon's work, but she wouldn't last another day before she'd be begging to climb in a chopper and fly back to the bank.

He wasn't far from wrong.

The *Condor* was on standby duty, tied to its home rig or a nearby buoy for weeks on end, one hundred and ten miles offshore, just barely rocking on the flat summer water. There would be no haphazard routine-breaking workboat come and go; instead we would rock, rock, rock in place. It wouldn't eat up more than half an hour of my day, tying and untying the lines when we moved from buoy to rig and back. The remaining eleven and a half hours of the workday would have to be devoted to chipping, grinding, painting. I might as well have had a factory job.

On that Day One of Hell Week, after suffering six hours of fire-and-brimstone sun on my winter-white skin, two of those hours spent hanging on to the mast with a gallon of oil-base paint in one hand, a slippery paint mitt on the other, I despaired in earnest. How would I survive six and a half more days like this one?

What saved me was a desperate impromptu attempt at malingering that turned into a full-scale epidemic. Woozy from the remorseless heat and sun, I rushed into the chill galley at noon, bolted down a plateful of hard round steak and rice, raced to the bathroom just across the hall, stuck a finger down my throat and regurgitated noisily: *ack gack gack garrrrh* for the benefit of Captain Billy, who could not fail to hear me.

Returning to the table with sick, sweaty circles under my eyes, I requested a little time in the cool of my bunk before I hit the deck again. "I think I'm coming down with something," I said. "It feels just like the time I caught amoebic dysentery from bad water . . ."

Bad water is easy to believe on standby boats, where the drinking water lies for weeks on end in scummy tanks below-decks. Captain Billy didn't doubt me.

By the time I woke up from my cool nap, the cook and the engineer had "caught" my disease. The mate decided that our freshwater tank must indeed be contaminated and barfed at the deck rail in the nauseated company of the other deckhand. By sundown Captain Billy himself was thoroughly sick. Our sister ship the *Heron* radioed that she had a similar epidemic aboard. My small ploy had exceeded my expectations. Billy radioed our office to have a relief crew sent out.

I could have arranged to go home then, but that was not my plan. With Captain Hell himself quitting the boat, I'd be home free for my trial week. So I made a miracle recovery, scrubbing out the grisly heads and moldy showers to prove it.

No plot is perfect; Byrd Marine couldn't find, on such short notice, replacements for the captain and mate. So I was stuck with Captain Billy and his little partner while the three other crew members made their getaway. The cook winked at me as the personnel basket lifted them away. "Hope I get well as fast as you did," he called to me.

"Sshhh!" I mimed at him.

But Captain Billy didn't suspect a thing. In fact, he eased up on me a little after that. I'd stood by him while the others deserted. So my phony epidemic paid off, and cut one day off my Hell Seven, giving my sunburn time to harden up.

Day Two began pleasantly enough. Captain Billy assigned his deck crew the job of scrubbing down the *Condor*'s hull from stem to stern. It would be cool work, at least.

Traditionally, the scrubbing job is divided between the two deckhands: one scrubs the port side, the other the starboard. The mate, when there is a mate, comes along behind, rinsing their work with a high-pressure fire hose. The job can take as little as three hours. A crew dedicated to wasting time peacefully will stretch it out to six.

But because Captain Billy stood watching us from the wheelhouse, I revved up my Wonder Woman act. For his benefit I took special pains to scrub the almost unreachable undersides of ladders and battery boxes. I scoured every davit on the lifeboats, cleared each drain hole in the bulwarks, every nook of the anchor winch. I kept my eye on Donnie, the replacement deckhand, to make sure he wasn't getting ahead of me. It would not do for Wonder Woman to come in a slow second, not on her trial week.

Whenever I glanced up at Captain Billy where he stood playing overseer behind the wheelhouse glass, I imagined he eyed me critically. I upped my speed. Soon the mate, a plump little sycophant and Billy's regular right-hand man, was having to whizz to keep up with me, struggling to haul that crack-the-whip fire hose from port to starboard and back. "Take it easy," he hissed.

"Is that Donnie kid ahead of me?"

"You're both going too fast."

"But is Donnie ahead?"

"Maybe a little. It's hard to tell."

I upped my speed another notch.

And so it went into high gear, a watery battle of the sexes, thirty-six-year-old adrenaline-powered Wonder Woman, the challenger, taking on nineteen-year-old Donnie Deadbeat, the Arkansas Strongarm Kid. No bookie would have given me even five-to-one odds, but I knew better. I knew the power of the adrenal cortex that whirred between my kidneys, pumping out waves of irrational go-power. Captain Billy had said I must "work right alongside the men." I took that to mean I must keep right up or get out. I would not fall behind. Telling my glands that this was my do-or-die moment, I surged ahead.

The competition was my fault; Donnie might have taken it easy if I'd done the same. It just didn't occur to me to go easy or even pace myself. I bit my lip and burrowed on down. Do or die.

By the time Donnie and I reached the opposite pair of stairs connecting the weather deck to the work deck, I'd called up my second wind, my third. Donnie just kept coming. I eyed his biceps, saw them bulging with effort, nearly faltered. But gads! He was one step ahead! I scrubbed down my stairs before he was halfway down his. Then he caught my action and slopped down to the deck in seconds flat.

Side by side, trading murderous looks, we scrubbed the galley's exterior wall. The mate was screaming at us now: "Wait a minute, for Christ's sake, I gotta switch hoses!" We bore on anyhow, snorting hard through flared nostrils, scrubbing our way down fifty feet of bulwarks and stanchions with a furious concentration. Then it was time to hit the tall smoke-blackened diesel stacks. Everything before this had been child's play, but we didn't slacken. Scrub!

The mate huffed far behind us, red faced and puny, wrestling with that kinghell hose. "Slow down, goddammit," he screeched. We didn't. We couldn't.

But I could feel the slump coming, the moment when I would flat run out of steam. A withering fatigue reached up my arms, echoing in the wind where my muscles used to be. All at once I couldn't go any farther. I had nothing left.

"Thaaat's better," the mate called out when he saw me fall back. Donnie wasn't falling back, though. One look at him rolling on ahead fired me up again. What the hell, just fifty more feet of bulwarks and stanchions. I'd come this far; I could finish.

Rounding the last turn of bulwark to the stern gate, I scrubbed so hard and fast I thought my heart, if not my brush, would burst. I was drawing even with Donnie, just barely. The mate trailed behind, zigzagging from port to starboard like an undercranked Chaplin.

Down to the last ten feet of bulwarks, and I pulled ahead. Donnie was drooping! Good. I poured on the oomph, inched even farther ahead. But Donnie must have seen me, must have had a foretaste of the shame he'd feel when a woman beat him out. Out of the corner of my eye I saw his scrubbing arm move in a furious blur as he pulled up even with me. We dropped our brushes at the same moment, at exactly the center post of the stern gate: photo finish!

I would have liked to sag to the deck sobbing, but the mate was scolding us, bug eyed and furious. According to his watch, we'd scrubbed the two-hundred-foot *Condor* in eighty-five minutes flat. Wow. I was impressed.

Captain Billy sauntered over then, crisp in his fresh bubblegum-blue jumpsuit, not a hair of his pompadour out of place.

"You missed a spot," he said.

If he'd said it to me, I would have thrown my bucket at his head and to hell with Hell Week. But he didn't say it to me. The starboard stack, on Donnie's side, was streaked with the faint gray film of diesel residue. My own stood proud and perfectly white. I was not just as good as the boy they'd pitted against me, I was better. The winner.

Winner of what, I asked myself later. Donnie and I might have lazed over that cool and pleasant job, killed half a day with it, put off the wracking labor of chipping, grinding, painting. We might have taken time out to sit in the shade and break a couple of bananas off the stalk in the cooler. We might have worked side by side and become, if not friends, at least working comrades. Instead we had declared war and exhausted ourselves early in the day. Day Two of Hell Week had just begun. Captain Billy sent us back to the bow to our real work.

Donnie, now he was no fool. He knew what to do about a hard-nosed captain: shirk. While I mixed the paint he picked for himself the job of painting the wheelhouse roof, out of Captain Billy's sight. "I done enough for one day," he said, and curled up for a nap.

That left me with the wheelhouse exterior to chip and paint. I chose the shady side and tackled the thick bubbly rust on the roof overhang. "I want to hear those chippers chipping," Captain Billy had said.

It was not quite eight o'clock in the morning and early in May besides, but the sun was already ferocious, radiating through a thin haze that did not mask but magnified its power to burn a human frame down to the bones. Sweat trickled over my scalp, down my ribs, collecting itchy bits of rust dust. Ten more hours, then five more days. I dug in and chipped with a vengeance, fired with desperation and dread. I knew I'd never last it out. I held off checking my watch until I believed an hour had passed; but no, it was 8:21. Time had slumped inert, doubled back on itself.

Inch by inch, with my arms hammering the chipper overhead, I chipped away the aching seconds and loathsome minutes until I felt safe checking my watch again: 8:49. A year later it read 10:15, eight hours to go. The sun's glare shot back from the bullet-gray Gulf waters and socked me in the eye even where I hid, in a narrow strip of shade. Smothering stillwater heat walled me in, made me wretched. I heard Donnie's comfortable snores and wished him a killing sunburn. For the captain, the mate, the engineer, who were loitering in the cool of the galley in easy reach of iced tea with fresh lemon, I wished more intricate tortures: grease burns, gout, cerebral-vascular accidents.

Five hours to go and I hated even the boat, everyone and everything on it. Closing my eyes against the sun glare, I met with throbbing yellow-white plasmoid shapes that caromed from edge to edge of a poppy-red ground. I hated them, too. For entertainment I wrenched up memories of old wrongs, ancient enmities, played them out and replayed them, imaging up old tormentors to dance as fools for my lurid imagination. Fat Jules I caused to parade pink-naked with a silver ribbon over his swollen belly, humiliated candidate for Baby of the Year. I nursed my hatred but time dragged on, tight, blistery, itchy as a new pair of boots on a twenty-mile hike.

Four and a half more hours. Scrap thoughts, junk worries, threaded through the ruins of my consciousness. Would I last out Hell Week? Why did I care?

I always say I've never been bored in my life. That's a lie. Like everyone else, I've pined away summer afternoons in doctors' offices, Laundromats. Once, when I was sixteen, I had a job stuffing junk mail into envelopes nine hours a day. So I've studied on boredom and I know that what makes boredom acutely painful is rage. I thrive on work, but being worked, and by a pinhead captain with delusions of being a seagoing plantation overseer, is enraging, is boring.

The endless day must have ended because I remember twisting in my bunk that night, radiating sunburn heat, panicking when sleep wouldn't come. I dreamed at last, feverish zigzags and stripes of nerve lightning, a dream not so much a dream as an abstract of my desperation.

Toward morning, a new dream came. This one was clear. Soft and simple with no sound and only muted color. A man was in it, a man I was given to understand was my dream guide. He would show me what I most needed to know, from a photograph in his wallet. The image in the photograph was a moving me, at work, chipping and painting the *Condor*. In this picture I wasn't struggling but smooth, peaceful as a rock at water's edge, my face clear and alive but without expression. My legs were folded under me in the half lotus, a Zen Buddhist *zazen* posture. The dream guide told me, without words but in a benevolent mind-to-mind transmission, what I had once known and since forgotten: that boredom is only a mirror. That I was only fighting myself. That all I had to do was paint, paint, paint.

I woke up from the dream and went out onto the deck. It was not yet dawn, and the rest of the crew was still sleeping. The boat rocked just barely on the smooth water while I mixed my paints. I heard the rig radio fizz a list of numbers in a broken voice. I heard the slap of an occasional wavelet on our hull. This was not the beginning of Day Three of Hell Week but a day I would possess for my own purposes. I was calling it *sesshin*.

I need to backtrack here, to 1971, if I am to explain *sesshin*. In 1971 many of my friends became Zen Buddhists. They were hot for Enlightenment. I was puzzled, especially when I heard a garbled account of one friend's visit to an American Zen meditation center. He had sat in place all day, every day, without talking, without seeing another human face or hearing a human voice. When he fell asleep, or wiggled on his sitting cushion, some Zen official would sneak up behind him and hit him with a stick. That's what I heard, and that's what my friends called *sesshin*. They thought it sounded wonderful; I thought they were crazy. Every day, in emulation of this traveler to *sesshin*, they sat *zazen*. Zen is not so much a belief system, and not at all a religion, my friends said, but a practice of sitting in one place until you come to Enlightenment as Buddha did. This Enlightenment might come in the first moment, or after years of painful *zazen;* there was no guarantee Enlightenment would come at all, but my friends didn't lean on that point. Crazy, I thought.

But my friends were doing it, and they were no slouches, so I tried this *zazen* once. I'd planned to sit, with my back straight and my legs folded under me, for the prescribed twenty minutes. I scrambled to my feet after five knowing now for a fact that my friends had lost their heads.

I told them I'd had "an interesting experience," but that was a lie. Nothing could have been more boring than five minutes of *zazen*.

My friends ignored me. I was weak, chickenhearted. They went on with their *zazen*. Eventually I tried it again. The full twenty minutes this time, and oh, the pain of twisted ankles, tense shoulders, wobbly spine. A great ugly bubble of wasted time, I thought. Surely Enlightenment, whatever that was, could not lie at the end of such a road.

I had never sat still for twenty straight minutes in my life. I'm one of those people who, when you see them alone in restaurants, are reading a book. When I watch TV I write letters and darn socks at the same time. I talk with my mouth full, light two cigarettes at once, and at bus stops I pace impatiently. I asked my friends what they were doing while they sat *zazen*. "We just sit," they said. Horrible thought.

This sitting, and I began to do it once a day, was truly unpleasant. After a while I saw what made it so. I was forcing myself to face the buzzing debris that was my conscious/unconscious mind: the large and small anxieties, the ignobly self-aggrandizing daydreams, the raw garble, gripe, and chatter of what I had been used to calling "my thoughts." From time to time, just sitting, I felt unbidden swellings in my genitals, sudden comings and as sudden goings of pains, pleasures, waves of irrational feeling. I would have given anything to escape this stuff, but this stuff was me. When I was up and about, going about the business of my "real life," I could only overlay the mindjunk with activity. I couldn't make it go away.

Famous Zen masters had written books about this experience. They said that after a while my scrap thoughts and random feelings would settle to the bottom of the barrel, or burn themselves away. Having now confronted this awful stuff of self, I couldn't live with it anymore. And so I sat, as the books instructed, simply paying attention.

After a year or two of off-again on-again *zazen* practice, I went to a *sesshin.* In Tibet, before the Reds came, Buddhists gathered for *sesshins* that lasted as long as eight years. The American *sesshin* that I attended was more modest: three and a half days of silent sitting, and for this American anyway, the ultimate spiritual endurance test. A year after that, I went to another, longer, *sesshin.* I hated *sesshin,* I loved *sesshin,* I hated it again. But it was all mine. Now, some years later, living in an alien culture, being worked like a slave by an alien pinhead, I reclaimed myself from Hell Week by declaring it my *sesshin.*

Chip, chip, chip. Then paint, paint, paint. It wasn't as easy as I'd hoped, settling into a private *sesshin.* I had four years of backlogged mindbabble to wrestle with, reestablished habits of blather and blither. But by afternoon of that first *sesshin* day I'd managed to clear my head a little. I disappeared into my work and the present moment for, at first, minutes, then half an hour at a time. Hot sun, sticky paint, intricate brushwork, *samadhi!*

I heard Donnie slapping occasional brushloads of paint in the circle around his rooftop sunning space, and I heard him bitch once or twice about his sunburn, but I pushed all that aside and painted on through. Paint, paint, paint.

When Captain Billy pulled a surprise inspection, I went on painting. I heard him dressing Donnie down, felt his pat on my shoulder. Then Billy was gone again and I continued to paint, paint, paint. The moments were sometimes as clear as clear water. Time did not drag but pulsed evenly.

Late that afternoon a helicopter came to the rig on a mission that surprised me: Donnie, and the cook who'd been caught with a bottle of whiskey in his bunk, were being banished. No replacements came. I didn't mind. I went back to my paint, paint, painting.

That evening Captain Billy told me that with the cook and deckhand gone, "each of us would pick up our share of the burden." My share was an extra hour of indoor housework and cleanup after meals. Sweep, sweep, sweep. Clean, clean, clean. I didn't mind; this was *sesshin*.

Which is not to say that I didn't resent the especially nasty test Captain Billy put me to.

Navy-style captains like Billy are famous for being very fussy about the indoor work, which they call "sanitary work." To make sure that I didn't skip scouring out every grundgy corner of the boat, Captain Billy devised the Toothpick Test, a piece of low-comedy Queegism that was hard to ignore, even on *sesshin*.

During the day while I was on deck, Billy hid twenty-four pastel-colored toothpicks one by one in cunning little spots inside the house of the boat. If I were cleaning thoroughly enough, I would find them all, one by one.

Because I was a Wonder Woman of a sanitary worker, the seek half of Billy's batshit game didn't give me much trouble. But on this third day I came up one toothpick short.

One little toothpick secreted somewhere in the three-story house of the boat. I doubled back on my tracks and searched. No toothpick. By counting the twenty-three toothpicks I'd already captured, I determined that the pick in question was turquoise. I went back over the boat again, tediously, in search of the elusive turquoise toothpick.

"Can't find that one blue toothpick, can you?" Billy interrupted me.

"No, sir, I haven't found it yet."

"Maybe I just happen to have it right here in my pocket," Billy said, producing it. "Just wanted to see if you're on your toes."

Days Four and Five went through me. I had, by the end of them, painted the entire wheelhouse and weather decks. Paint, paint, paint.

I overheard the oil company man, down from the rig at Billy's invitation to a "gormay luncheon," say he admired the work I'd done so far. "It's a regular shipyard finish," he marveled. "And her just a woman."

"She may be a woman but she sure can work," Billy told him. "I wouldn't trade her for any two man deckhands this company's got."

I tried to tell myself I was doing this work for its own sake, part of my *sesshin.* But Billy's praise ruffled my work trance, I admit it.

Billy was so proud of me he even let me clean my brushes fifteen minutes early that evening. Free time, my first free time of the week. I made the mistake of spending it in the galley, where Billy labored over another "gormay dinner." I discovered how lucky I'd been to be banished to the deck all week. Billy's bullshit was extravagant, embarrassing.

Supper that night was another assault on round steak ("It's the bay leaf and cinnamum makes it gormay," Billy said) and Captain Billy claimed that his recipe for it had been "plublished" by the "New York Gormay Sociation."

After dinner, when it was time for my sanitary work, Billy exempted me from that too and invited me on deck for what he said would be a lesson in astronomy. That sounded good, but the lesson degenerated almost immediately into a silly soliloquy on Billy's loneliness at the top. Ah, the moon. Ah, the stars. Ah, a captain's life was hard and lonesome with only the night sky to tell his troubles to.

I caught a whiff of courtship in the air and changed the subject back to stars. Wasn't that Sagittarius up there carving an arc in the sky?

Billy didn't know, and deflected my attention from his astronomical shortcoming by telling me he'd shot the Loch Ness monster on just such a night as this.

I shammed surprise. "I didn't know you'd actually been to Scotland."

No, Billy claimed he'd bagged old Nessie on one of her rare visits to the mysterious Bermuda Triangle. He'd positively identified her as a "rogue Blazosauras," and been written up in "all the scientific-type magazines" as a result. "I forget all their names right now," he said.

On the strength of my apparent gullibility and his scientific-type credentials, Billy went on to inform me that the Egyptians had come from outer space to discover America by true celestial navigation. He knew it "for a fack."

And no, sadly enough, this poor old buffoon would not admit to pulling my leg. When I suggested he might be teasing me, he sent me back to the house of the boat to polish his boots. Polish, polish, polish.

Day Six and my paint, paint, painting was nearly seamless all morning long. I didn't bother to move with the shade anymore. Come on, sun, do your worst. Pour it on!

Captain Billy interrupted me twice, still stinking of eau de woo. While I painted on, he stood over me in proud captainly fashion reading prurient passages from his studfarm slave novels, paragraphs dealing with the weight of dusky testicles in aristocratic white palms and sweaty black bucks writhing on the auction block. Billy seemed pathetically eager for me to make some response. I made polite noises and went on with my painting.

Trouble is, a boat is a small place and there was no real

escape from Captain Billy's swampy subconscious. He had plans for me. Over lunch he told me that since I was doing such an excellent job of renovating the *Condor* he'd decided to take me with him to work on his regular boat, the *Sandpiper*, flagship of the Byrd Marine fleet.

Caught unprepared, I cried, too loudly, *"Oh, no!"*, followed up too late with a too flustered demurrer. "I mean, I'd want to get lots more practice with the lines, and ah, really, I appreciate what an honor that would be, but . . ."

But by that time Captain Billy had clamped his jaw down so tight on his dentures that he just about swallowed his nose. Billy's faithful sidekick, the mate, shot me a how-could-you look.

That was only the first snag in my day. The second came late in the afternoon when the rig called us to come over there and take on a load of scrap machinery.

The roustabout the rig sent down to help me handle the crane work was an eighteen-year-old black man. This was his first day on the rig, his first time on any boat anywhere. The *Condor* was rolling pretty hard that day and he wasn't on her deck five minutes before he'd puked in his boots.

I brought the deck hose so he could wash himself, and furnished him with some soda crackers from the galley, a specific for seasickness. He wanted to know if the boat was going to fall over. I told him no and sat beside him on the skid while we waited for the crane to start its work. We talked a little bit. He was my daughters' age; I tried to reassure him about boats.

Then Captain Billy's voice boomed over the deck's p.a.: "The deckhand will report to the master of the vessel. Immediately."

I scrambled up the ladders to the wheelhouse with my heart in my throat. His voice told me I'd been caught. At what?

"Thought you'd want to see the report card I made out on you." Billy handed me a yellow slip of paper, a standard evaluation form for deckhands that would be returned to Byrd Marine's office at crew change time. A for efficiency, it said. A for reliability. B for attitude.

"You want to know whyyy I gave you a low grade for attitude? Because you wiggled your butt at that nigger, that's why."

I blew up, but stupidly. I didn't take time to sort out his one delusion from the other. Glib just failed me; shrieks and stutterings were all I could manage.

But I'd said enough. Billy sent the engineer to the deck in my place and banished me to the engine room companionway bilges, to the corner where the shit tank (Billy called it the sanitary tank) stood. He wanted that area scrubbed "so clean we can eat off it" by dawn, when the crew changed.

"I'll put you doing the nigger work on this boat if you're such a damned bleeding-heart niggerlover," he said.

Scrub, scrub, scrub, but the hours dragged this time. My detachment was destroyed. B for attitude.

I got through the night by reminding myself that I was in the last hours of my audition for the deckhand job. I'd make it, by God. Still, I had to promise myself that I'd someday get my revenge.

I got it, too, sooner than I'd hoped.

When I climbed out of the sanitary tank bilges to report that they were spotless, I found Captain Billy alone in the wheelhouse, bringing the *Condor* out of the Gulf to the Freshwater Bayou. But something twitched my navigation nerve; our position was not right.

Captain Billy, who taught radar, loran, and celestial navigation at the local marine academy, had lost our course. He looked as if he was getting ready to cry.

"The compass is broken," he raged, pounding it with his fist. "Completely off. I don't know where the hell we are."

I recognized our position from a clump of onshore towers, having made the same navigational error myself back on the *Harbor Pride*. Captain Billy had overshot the sea buoy by fifteen miles.

He must have seen some evidence on my face of the superiority I felt. "I don't suppose *you* know where we are," he said, suspiciously.

"Actually, sir, I believe we're about fifteen miles east of the sea buoy and into Ship Shoals."

"And what makes you think *you* know anything about that?"

"Just guessing, sir."

Billy wouldn't take my word for it, of course. He ran the *Condor* up to a pogy boat that was tied off to a well cap and blew our air horns, startling the pogy boat's sleeping crew out onto their deck. He hailed them for a compass reading and position. They hailed him back. I stole a look at our own compass. Perfect agreement. Billy traced our position on the chart with a shaky finger. I'd been right. We were fifteen miles east of the sea buoy.

Billy only snorted, and manhandled the *Condor* onto course. An hour later I spotted our buoy.

"We ought to be just about there." Billy squinted. Damn! The man couldn't even see the sea buoy! I could have pointed it out, but hell, he'd made his masterly bed; let him lie in it.

Guste had taught me the tricks for coming across the shoals into Freshwater Bayou Locks. It's fairly chancy as these things go, what with swift local currents and a squatty black sea buoy that's hard to spot in daytime, when it isn't blinking. Guste had told me that sooner or later I'd run with a green captain and get a chance to save his ass by bringing the boat in myself, so he'd let me practice piloting that stretch again and again.

But Billy didn't ask me the time of day. He rang the *Condor*'s general alarm, summoning our mate and engineer.

A general alarm is an exquisitely alarming sound, designed to scare the living bejeezus out of everyone aboard. The two men popped up the stairs just moments later, wide-eyed with panic, wearing their silly jockey shorts. "Find me the sea buoy," Captain Billy commanded them. "And *you*"—he indicated me—"keep your damned mouth shut."

The *Condor* rocked for long minutes on the spot, her wheels kicking up Louisiana mud while the two crewmen fuddled around rubbing their eyes. I examined my fingernails. Damned if I'd tell them where it was.

The porpoises whose purpose in life seemed to be leading lost boats to the Bayou Locks hung in the water off our bow looking puzzled. Why weren't we coming in?

Finally the captain turned to me. "I'm willing to stake my professional reputation that somebody sunk the sea buoy," he said.

I said, "Captain, that's the sea buoy right over there."

With a savage wrench of his arms Captain Billy wheeled the *Condor* around to take the buoy on our starboard side. The wrong side. Even at high tide and in smooth weather, the boat must be lined up just so between a cluster of offshore platforms and a big white chemical tank onshore. Then the captain must squeak past the sea buoy on the port side full speed ahead or the currents will whip the boat around and strand it on the mud.

For expediency's sake, the helmsman can just follow the porpoises. They know the way.

Today the seas were rough, we were coming in at low tide's turning with the currents especially nasty, and Captain Billy ignored the porpoises altogether. We were about to run aground.

Billy must have spotted my white knuckles clutching the starboard rail because he turned to me again. "What's *your* problem, Miss Liberal Bleeding Heart, Miss Think-Yew-Know-Everything?"

"Sir, I believe it's wiser to take the buoy on our port side."

Billy sent me to my quarters then, and I was no more than halfway down the outside stairs when I had the satisfaction of feeling the *Condor* run aground with a hard *scrump*. A sea gull landed just a few feet from our hull and took a walk around us, pacing as if he were worried. Pacing in three inches of water.

I heard the bow thruster kick in with an impotent wailing

whine. Nope, we were stuck good. Beached. I went on down to the galley and compiled a lusty triple-decker sandwich, a super-Dagwood garnished with my own delighted smirk. Maybe I would plublish it with the New York Gormay Sociation. It certainly was tasty, I thought, listening to Billy's dozen more screeing, scrowling attempts to back the *Condor* off the mud. Then the diesels balked into silence. Billy must have stuffed up our exhaust ports with blue gumbo mud. That called for a fat slice of chocolate cake. Revenge is sweet.

An hour later I was touching up the paint on deck when Guste arrived with the *Pride* to tow us off the bank. I waved to him. He waved back, rolling his eyes. Both crews strained for an hour to extricate the *Condor*. When this was done, I heard Guste's half of a radio conversation with Billy. "Yo' got dat woman deckhan' dair, give dat wheel to her. Ah trained her up mahseff, me, an' dat gel can tear yo' a new asshole comin' in fo' dis here bayou." I waved again. He waved back, and headed on offshore. I hated to see him go; now I wouldn't get to visit him when we hit Intracoastal City. Still, he'd shared in my revenge. He'd know.

Even when Captain Billy followed Guste's instructions to take the buoy on the port side, Billy nearly cut a shrimp boat in two with our bow. I realized then what I should have known from the first: Captain Billy was crippled with nearsightedness and too vain to wear glasses. Being a retired Guardsman, he'd probably winked his way through the Coast Guard vision test when his license came up for renewal. I wondered how much longer he'd be allowed to play captain, how many more crews would be endangered by his vanity.

But hell, I could forgive him. I'd heard the company radio confirm that I'd passed my one-week trial. Captain Billy would be leaving the *Condor* and I'd be staying aboard as deckhand, a tried and proven ordinary seaman on a full-sized supply ship. And no matter what sort of beast they sent me for a captain next, I knew I'd get by, now that I had my paint, paint, paint. I had, I believed, found the way.

17

Just after dark that same night, at dock, with the taste of rank and blooming earth hanging on an occasional warm ruffle of wind, Captain Billy and the mate sneaked away home. Coast Guard regulations forbid any merchant mariner, most particularly the captain, to quit a vessel before his replacement arrives. As usual in the oilfields, regulations were ignored.

The mate took me aside to coach me in a story: When the new captain came aboard, presumably at dawn, I was to say that Billy had departed just five minutes before on urgent business. "He's a good man, really," the mate apologized, and slipped ashore.

I didn't mind their leaving. Somewhere belowdecks the only other crewman aboard, our pitiful excuse for an engineer, was busy scrubbing down the engine room, making up for his week's time lost at Billy's ongoing Mad Hatter tea party. For the night, the *Condor* was mine.

Not such a bad boat, I thought, not at night when shadows softened her glaring whiteness and ungainly proportions. I washed down her new paint, gently, with the dock's fresh water, and laid her lines just so on the bitts. I could feel my shoulders loosening in their sockets, relaxing out of *sesshin* and striving.

I was feeling wishful, too. I hadn't yet overcome the habit of hoping for a good crew, a warm welcome aboard. If this is hard for you to imagine, imagine this: On your first day at a racetrack, by a fluke in which you see the hand of divine justice at work, your ticket on the first race pays off forty to one. Someone that you may be and I am not will take the money and run. I'd still be there at the betting window at the bitter end, and the next day, too. So although I didn't expect my seventh captain to be another Guste, I hadn't lost faith in divine justice. I aired out the staterooms and made up the bunks with stiff white sheets, then baked a pair of fancy cinnamon breads, a welcome.

At dawn I stood with the engineer in the wheelhouse, ready. The engineer, whom I had nicknamed Kewpie, appeared to be as wishful as I was, and maybe more vulnerable. Just twenty years old, Kewpie was plump and pink as babies' feet. I would have guessed, just looking at him, that he'd relied all his life on his baby cuteness. I would have been right. His professional-looking engineer's toolbox contained no wrenches or screwdrivers, no socket set, not a single tool of any kind. If the *Condor* broke down, we were helpless. Kewpie's toolbox was packed with eight-track tapes. "Kiss an Angel Good Morning," "Harper Valley P.T.A.," and, because maybe somebody up there did like me, Bruce Springsteen's "Born to Run."

All week Kewpie had batted his baby-of-the-family blue eyes at Captain Billy's ever more outrageous Loch Ness lies, laughing behind his hand as if to say, "Yeah, I'm some kind of suckass but at least I know it." I didn't like the kid at all. I even had a special personal grudge against him.

Just minutes after Kewpie came aboard I'd been slinging the rig's lines off our stern bitts, a rapid-fire high-tension job, when he shouldered me out of his way and ripped the lines from my hands.

"Watch out, lady," he said. "I'll handle this."

I shoved him back and thundered in my best sailor's English, "Fuck you in the mouth, kiddo. This job is mine."

Above the roar of our main engines I'd heard Kewpie's

plump little feet pounding across the deck and up the ladders to where Captain Billy stood at the stern controls. "Cap'n! Cap'n! She said to fuck me in the mouth!"

Captain Billy took me aside later to advise me, not unkindly and all too sincerely, of the importance of my remaining a lady even while attempting a man's job. Kewpie and I didn't speak after that.

But on this crew-change morning when anything might happen, we talked. Kewpie said he had to admit he hadn't believed any woman could work the deck. Now he thought maybe one could, "if she was a certain kind of woman, and if everybody watches out for her." He asked what I was doing on the boats; I looked more to him like a college graduate type or a mother, something like that.

Sliding out from under that mild insult, I turned it back on him. Kewpie didn't look like a standard model of sailor, either. What was he doing there?

"Oh, I don't know, really. My family owns this big real estate company back home in Tupelo and they give me a good job, too. I guess I just needed to get away and prove myself."

What was the good job he'd left behind?

"Well," Kewpie said, leaning closer, suddenly grown-up and businesslike, "what I got to do is kick the niggers out of the rental houses where they more or less wormed their way in during Civil Rights."

Dig it. As if civil rights were not an ongoing constitutional guarantee but an unfortunate bit of past history, like the Civil War. I'd heard this kind of blind blather from airheads all over the oilfields, and it nearly always set me off. But this morning was too warm, too easy, too fair. Even the dingy bayou sparkled pink and gold under the sunrise. I moved away, down to the deck, to inspect my lines one more time. Everything had to be shipshape for my new captain.

So that's what I was doing when the two men, thick as trees and drunk besides, came aboard. They were biting and growling, scuffling up the dock's ramp to the *Condor*, making a god-awful racket and bumpy progress. Behind them, dissociating herself from them, walked a patrician-looking young woman, tall, long in the neck, with a fizz of russet-brown hair and a disdainful row of wrinkles on her nose. Who could this strange bunch be? I walked over to meet them.

The men stopped wrestling, as if on cue, when they reached the edge of our deck. The larger man lifted the smaller into his hammy arms, as a groom would a bride, and crossed onto the boat. I looked back to the young woman for an explanation. She only widened her nostrils into deeper disdain. The men were wrestling again, thumping across the deck to the opposite gate where the man who'd played bride dumped his groom over the side and into the bayou.

The big man bobbed to the surface, spitting and pawing the bayou garbage out of his face. "And all because I said your tits wasn't big enough! You a mean old bitch, you are."

I remember wishing when I was five years old that someday I might have a chance to save a life. Here it was: man overboard, filthy bayou, hard current. I went for a line and a life preserver but I didn't get far with them. The smaller of the two strangers leaped on my back and bit me on the neck, hard. "Don't even know your own cap'n 'less he come up and bite you."

He introduced himself as Mick, Captain Mick. "And don't go gettin' you a shot for lockjaw or nothin' 'cause I ain't got no teeth nohow, har har har." He yawned clownishly to show me his bare gums. Hard gums, though. That bite had hurt.

The man in the water, swimming strongly now for the marsh island one hundred yards away, was Gitzy, short for Gisclair, the *Condor*'s mate. "And that there"—Mick indicated the young woman—"is the cook, Karen. She's some kinda comp'ny spy. Wouldn't drink a drop all the way out here. I got drunk as a whore myself and she can tell the whole fuckin' world about it 'cause, frankly, my dear"—he aimed this at her—"I don't give a flyin' fuck."

Mick extended a hairy paw to Kewpie. "This here the engineer?"

Kewpie offered his own pink hand, eager to please but uncertain of how to go about it. Mick wrung it, then twisted his wrist in some kind of homemade judo move that brought Kewpie to his knees on the deck.

"Hee-yaah!" Mick said, then let Kewpie go again. "Don't look to me like you're wuth a shit, boy, but that don't matter. None of us is wuth a shit or we wouldn't be on this here old scow, now would we? Har har har."

Clasping my own hands safely behind my back, I identified myself as one of the deckhands and told Captain Mick I hoped he'd brought another.

"I noooo I forgot somethin'," he howled, walloping a palm into his forehead so hard he knocked himself off balance, and led me back to the company's carryall van. Inside it a tousle-haired cutie sat smiling dimly, angelically, unseeing, into the floorboards. "That there is what we call SpaceMan," Mick said. "Can't handle his liquor but other'n that he ain't wuth a shit. SpaceMan! We is arrived! All aboard!"

The man called SpaceMan just giggled, and drooled a little. Mick grappled him under the arms, I caught his feet, and we carried him aboard.

"Tooth head," SpaceMan slurred, then corrected himself. "Take me to the head." We deposited him on the toilet, upright and fully clothed. He passed out cold.

Karen, the cook, threw us a dirty look from the galley.

I hate to say it, but I liked them already, all of them. They do sound horrible, I'm sure, but they were real merchant mariners, a motley human crew and some of them hundred-proof heathens besides. I liked them just fine.

Mick, who could see that I did, hugged me up under one arm and carried me back to the deck to see what had happened to Gitzy. Gitzy had made it to the marsh island and was lying on its muddy edge with his gargantuan belly up, a beached whale of a man. I saw him grin, and wink. Mick cried, "To the rescue!", and set off with Kewpie in the lifeboat. "Miz Wucy," he shot back at me, "go see if you can round up a little drop of somethin'. That pore shipwrecked sailor's gonna need a re-storative."

By the time I was back on deck with the last smidgen of booze aboard, Gitzy and Mick were rowing the lifeboat back across the bayou. The Owl and the Pussycat, they looked like. They'd left Kewpie on the island, pink and forlorn. Did they want me to row back for him?

"Shit no," Mick spat. "Little suckass, ain't he? I can smell the skunk on him already. One spy is plenty. Leave him out there for the alleygators."

Mick and Gitzy ransacked the walk-in freezer next, hopping from bare foot to bare foot on the cold floorboards, looking for fish bait. "Looky here, Gitz, we got ten pounds of big old brown shrimps. Snappers love them brown shrimps."

Goodbye shrimp gumbo.

"And beef roasts, we got any beef roasts?" Gitzy, still hopping, hurled hard white freezer packages over his shoulder. "Can't get sharks wiffout beef roasts," he swore. "Attaboy, here's one. Nice one. Ever catch you a big old shark, Hoss?"

I told him I'd never caught a fish in my life. He promised to teach me. "And when we get us a shark? We'll cut his eyes out of the sockets and boil them up on the stove all night till they turn hard as stone. I'll drill me a hole through 'em and make you the purtiest macramé you ever seen with the eyes for beads. Got to save the jaws for my boy, though. He likes the jaws. We got jaws all over the house."

Mick sent me off to the liquor store with a handful of loose bills to buy up all the cheap booze I could get for that amount. When I came back with six quarts of bourbon and vodka and some small change in cheap wine, Kewpie was aboard again, sulking at the stern, SpaceMan hadn't left his toilet throne, and Karen, still miffed, was rearranging the galley to her own taste, working around Mick, who sprawled asleep on the galley table, snoring like a jackhammer. She called him, under her breath, "that filthy clown."

God, it was wonderful, having another woman aboard.

I heard the whine of the diesels cranking up and rushed on deck to loose our lines. We were taking her out to the Gulf now, "just you and me, Hoss," Gitzy shouted, "all by our-seffs."

No matter how many times I'd been aboard a ship gunning out of dock to the Gulf, I hadn't worn out the thrill: the *zwheeeeng* of diesels cranking, the rumble and cloud of black smoke when they caught, the whip of lines coming off the bitts, the surge of power that shivered down the hull. We wouldn't see land again for twenty-one days. That's what I called heaven.

Gitzy at the wheel was a comforting sight with his naked feet splayed out on the floor, planted firmly. His center of gravity rolled easy in the massive belly that bloomed over the waist of his raggedy shorts. Unlike Captain Billy, who navigated, and so poorly, by instruments, Gitzy hadn't switched on the radar for this daylight run. He piloted with a ballsy serenity on the evidence of his own senses, meaty hands riding lightly on the wheel, going, I knew, by the feel of the boat in the water. A true seaman. Gitzy even matched my original picture of a sailor: a mess of curly red hair under a faded cap, untrimmed red-orange beard, juvey hall tattoos on his massive forearms, gold ring in one ear, and the worst case of permanently bloodshot eyes in all the Gulf of Mexico. (Later I heard Mick tell him, "Gitzy, you better close your eyes before you bleed to death.")

What's more, Gitzy was smoking a joint. He offered it to me. I asked him if he worried about warping his perceptions when he had a tricky stretch of bayou to pilot.

"Hoss, I was stoned when I learned it. Don't believe I could drive it now 'less I was at least a little bit messed up."

There are people I feel easy with, and people I don't. There are men who when they touch me radiate, if not a lust for flesh on flesh, at least a lust for the appearance of ownership. But when Gitzy brought me under his arm that day I felt the uncomplicated comfort I'd had with Guste. Actually I shivered a little at first; I'd been too long unhugged. Then Gitzy's ease crept over me and I relaxed. We would be friends. We would all be friends.

The two of us swayed together at the wheel for a while, sharing the joint, watching the *Condor*'s bow push up a steeple of water, watching our wheelwash suck at the bayou banks. Then Gitzy stepped back and left me at the wheel. We were coming up behind a slow tug and a pair of barges that blocked the narrow bayou. A tricky pass complete with proper horn signals must be executed. I managed it without blinking; it seemed so right, being back on the wheel of a great mother boat, on the waterway I knew best. Gitzy patted my back, apparently satisfied with my seamanship, and left me alone in the wheelhouse to take the *Condor* to the Gulf.

Hell Week was over and I had it all again, the reward for every pain and danger, trial, humiliation, defeat: a big old boat under my hands and sailors, true sailors, for companions.

True sailors are, by my definition, scattered over the earth, most often lost from the sea. They are the mad dogs and lone coyotes, most often male, who hole up in cheap boarding-houses, corporations, huts at the far edges of our villages, drinking their pain. The pain, if they could stand to feel it, is only a question. The question, if they had the wherewithal to frame it, is anyway cosmic, ineffable, unanswerable, and drowned by a roaring in their ears. A man of this kind may brace himself against the roar to live and die confounded by the question, knowing only the blurry edge of the pain of it. Or he may someday suddenly refuse the pain and find himself carried by the roar in his ears into the streets of Earth, suddenly so nonchalantly shooting nineteen strangers stone dead in the street.

Some few lone coyotes, the lucky ones, discover the sea, recognize the roar of it, cast off their confoundment, and do what sailors must do to redeem themselves: ride the roaring waters, run.

By no means is every seaman a sailor of this kind. Even when true sailors meet on the sea, they don't often extend their meeting into friendship, society. Only once in an ocean of great whiles do they strike up a reckless camaraderie, the way strangers walking the streets after a hurricane or a riot may do, roaming together, looting the ruins with their eyes.

Mick and Gitzy were two such men, and here I would be running at their side for a while. I didn't consider that they could be dangerous to me; I approved of them and they knew it.

Karen, when she joined me in the wheelhouse, was not so approving. She said SpaceMan had thrown up and was *lying in it.* Mick and Gitzy were sleeping on the galley benches, right in the middle of everything, *with their arms around each other.* If I hadn't been a Yankee like herself, she probably would have extended her distaste for this crew to me.

Karen had turned thirty just the day before, and this was the first day of her own midlife adventure. Hearing her history, I would have thought she'd adventured enough. She'd lived in every major posthippie culture center in the United States and hitchhiked through the Far East besides. She'd worked as a macrobiotic cook, an occult bookstore clerk, a holistic masseuse, a yoga teacher, a cocktail waitress, a California fire ranger. Now she was running away to sea at the invitation of her uncle, an executive with Byrd Marine. "It's probably got something to do with Neptune transiting my natal Midheaven," she mused.

We came to like each other quickly. I showed her how to plot a course, and she plotted ours easily. When we came into the Bayou Locks she tied and untied the *Condor* as if she'd been born on a deck. She stood behind me while I piloted, wrenching the tension out of my shoulders with her holistic masseuse's strong hands. We fantasized about our someday all-female crew, imagined the five of us women reading Proust and (her suggestion) Gibran in the evenings at dock.

Out on the Gulf, Karen took the wheel and played with the boat as I had done when I was new to the *Pride.* I sacked out on the bench behind her feeling perfectly satisfied with my life. The *Condor* plowed on pleasantly through easy seas and the salt breeze drove out the last heavy breath of land. This was the stuff; this was what I'd come for.

After a while I heard wet feet shuffle up the outside stairs.

"She's kinda cute, lying there," I heard SpaceMan say. Karen, still disgusted with the male members of the crew, sighed theatrically and dodged up to the wheelhouse roof to sun herself. SpaceMan and I had some hours of conversation.

SpaceMan was the first of this new crew to grill me about the suitability of women to work on deck. "How come you want to play deckhand?" he asked.

I was smug, stoned, and comfortable enough that day to roll out a speech I'd been rehearsing, about how it didn't seem fair that when boys grow up they need never give up their toys. They could go on playing with bigger and better ones: workboats, cranes, forklifts, helicopters, while the girls stayed home, getting serious. What the men kept calling the Man's World, as if it were some new variety of theme park, was just that: an amusement center for the terminally boyish and, in my case, the terminally tomboyish. I asked SpaceMan if he would turn in the thrill of lassoing high pilings from a moving deck to go home and fool around with real-life dolls and dollhouses.

"Well, no," he said. "But that's different."

"SpaceMan," I told him, "other women will catch on to this sooner or later. They're bound to. I'm just the wave of the future." At that moment I felt like it, too.

"You make yourself too complex," he said. But he said it almost kindly, admiringly. He sure was cute.

Watching him at the wheel, draped over the pilot's seat, the way his bones were hinged so-loosely, the way his dark brows and messy beard and studious no-nonsense spectacles made him look like a young Arthur Miller, I could feel a crush coming on.

Right feeling, wrong man. Over another joint SpaceMan let spill his own preoccupations. "I'm better off smoking dope than drinking. I've been having these attacks." He claimed he couldn't remember what the doctors called it, but I recognized the affliction from his description: sudden outbreaks of unconscious violence, true blackouts from which he emerged, groggy, moments or hours later, to find friends and bystanders smashed up and bloody, cowering from him. Jacksonian epilepsy, I guessed. Van Gogh had had the same trouble, I told him.

"Van who?"

"Van Gogh, the famous painter."

"I got mine from a motorcycle accident," SpaceMan said, parting his hair to show a livid purple seam in his skull.

He asked me, by the bye, what had happened to "that little pinky guy, the one who's supposed to be the engineer."

I told him how Mick and Gitzy had marooned Kewpie on the marsh island that morning.

"That accounts for the bug bites all over him, but how'd he get that purple goo in his hair?"

That purple goo, I found out later, was blueberry pie filling Gitzy had dumped over Kewpie's head as the boy slept in the sun of the back deck. The party was starting. And Kewpie's Hell Week was the life of it.

18

I hate bullies and bullying. Whose mother doesn't? But there in the oilfields, in what anthropologists call a men's house culture, bullying is the sport of kings. As Mick tried to teach me, "Top dog fucks the bottom dog. That's the law of the jungle, Wucy." Mick spoke, of course, in his capacity as king of that jungle. Gitzy was his cruelly inventive sidekick. Kewpie, our youngest, our weakest will, was the cur dog at the bottom of the heap. His mama's boy life strategy had served him well with Captain Billy; now it made his victimization all too easy.

I'm tempted to say that Kewpie was a natural for the part, but his sufferings were real, painful to watch. More on that later. In the confines of a small ship anchored one hundred and ten miles from civilization, the top dogs had the bottom dog cornered and too much time on their hands.

First came the gooey substance attacks: raw eggs in his bed, a shaving-cream pie in his face, oyster soup in his boots. When Karen cut off the top dogs' supply of gooey substances, they left a stew of human excrement in his toolbox and moved on to the left-handed monkey wrench series of pranks. Kewpie took too long to catch on. He spent one whole morning in search of the rudder clinch screws, another afternoon tracing a lost three-gang pulley. And damn him, he never got mad about it.

Then came gophering. "Go fetch my tackle box," Gitzy commanded him. Kewpie fetched it. Engineers are, in merchant marine hierarchy, equal in rank if not superior to ship's mates. When Kewpie balked a little at the second gophering run, Gitzy leaned over the galley table and pinched Kewpie's nose between two fingers. "You fuck with me, boy, and you'll be unzippin' your pants to brush your teeth, knock you so hard on that purty head of yours."

SpaceMan stood a little apart from the active hazing, but he laughed on cue. Karen at first held herself aloof, but as her friendship with Gitzy grew, she was chuckling with the rest. Kewpie alternated between false aplomb and a too-pink eagerness to be played the fool.

As for me, that is the sticky part: I was a sport. Served the suckass right, I thought, him and his "Cap'n! Cap'n!," him and his Ku Kluxisms, his loathesome batty blue eyes. I'm ashamed to say that I enjoyed his tortures. Is the army like this? Are we all so easily corrupted?

No, I'll take it squarely on my shoulders: During the gooey substance phase of the ganging up, I volunteered to sneak into Kewpie's bunk room while he slept and drop his hand into a bucket of warm water. Practical joke theory says this trick brings on bed-wetting. I never found out because I stopped myself, just in time. I'd been a bed wetter myself when I was a kid. I came close to identifying with the bad boys, but not that close. All the same I had no sympathy for victims, cur dogs, laughingstocks, not at first. I was a sport, grateful to be one of the gang I loved.

There were the small pleasures: the loose, everyday fact of our physical ease with one another; I was getting all the companionable hugs I needed for maybe the first time in my life. I remember Mick standing behind me brushing my hair, running his motherly fingers through it as if it were water, while I played slow, thoughtful games of backgammon with

SpaceMan and Gitzy taught Karen the art of macramé. I remember a particular Sunday dinner, fried chicken so delectable we sucked it down to the bones and composed an impromptu and goofy song about it, so delighted with our new friendships that we played a five-way game of barefoot footsie under the table.

There were small individual companionships, as when Gitzy taught me how to tie a sheepshank and a cat's-paw and that ultimate of knots, the monkey fist. And once when SpaceMan and I rummaged through the ship's stores we uncovered a huge cache of outdated emergency flares. That night even Kewpie was part of our fireworks party and we decorated the dark night with blue, green, red, and yellow cloudlets.

Mainly, for me, there were the times when Mick allowed me to play skipper.

When I was ten years old, maybe only nine, my father— who was probably drunk at the time—asked me if there was some one thing he could teach me how to do, some grown-up thing that would please me to know. Our family was vacationing in Florida, and when he asked me that question we were just the two of us driving a sandy back road into town to pick up some groceries. I said I wanted to learn to drive a car, never dreaming he would allow me that greatest of grown-up pleasures. But he did. Since my feet wouldn't reach the pedals, he did that part, but he allowed me to steer the great steel monster and I actually ran it off the road a couple of times before I got the hang of it. Sitting on my father's lap on that hot and sunny day, I even learned how to shift the gears and blink the turn signals. It was the thrill of thrills for a pigtailed child age nine or ten.

And you know how the things of childhood—grandma's backyard apple tree, that kind of thing—seem to have shrunk when you meet them again thirty years later? This recaptured thrill of my childhood had not shrunk but expanded; the *Condor* was no 1950 DeSoto but a two-hundred-foot-long, three-story-tall ship of the seas, and Mick allowed me to play at its wheel as if it were my very own toy. Expensive toy, too: the *Condor* burned one hundred gallons of diesel fuel per hour, and these sessions of playing skipper lasted at least that long every day.

Mick let me do it all. I ducked down to the engine room and cranked the diesels, ran forward and threw off the buoy line, then took the helm and ran her away from the buoy into open sea. I discovered, through play, just exactly how tight a circle she could turn, how fast or slow she would lose what is called her way when I cut power, how to nestle her in next to the buoy again under varying wind and current conditions. Then Mick said he would teach me to dock her at the rig.

Docking at a rig is many times more difficult than docking at a dock; a rig is at sea, not sheltered in a harbor, so the boat is at the mercy of wind and current from any and all quarters. No one knew how to dock a boat at a rig until sometime around 1954 when the first offshore rig was jacked in place. Then the boat crews and the rig crews worked it out between them, by trial and error, until methods were found. None of those methods is easy, or foolproof. Just laying anchor is a pesky enough problem with unmarked underwater high-pressure pipelines nearby, but laying anchor in a trackless sea at just the spot to allow the boat proper scope to butt her stern against the rig's crane side is a tricky, tricky piece of business. Trickier yet is for the helmsman to hold the boat in place while the rig's crane crew and the boat's deckhands mess with the lines.

Mick spent long hours teaching me to do it, first in open water with makeshift milk container buoys standing in for the rig, then at the rig itself. I was, I thought, a painfully slow learner, but Mick didn't seem to tire of my trying. Like my father, Mick was probably drunk. All Mick said about it was that someday, when I was a captain, he'd be able to brag that he'd taught me all I knew.

Eventually I got it right. It took me almost as long to get it right a second time. Mick had me practice it a third, and a fourth, before he was satisfied. Then he slapped me on the back and said he was proud of his Miz Wucy.

I sure did love my crew.

None of us did much work; the sea play and the friendly patter were too good to miss. I remember sitting spellbound when Mick and Gitzy told their sea stories. (According to Mick, the difference between a fairy tale and a sea story: "Fairy tale starts out 'Once upon a time . . .' Sea story starts out 'Now this ain't no bullshit . . .' ") Mick was a grand master of the bad joke. Gitzy was something more, a natural-born true-life storyteller. He had a pocket-rocket barroom rap that wouldn't quit and his scarred clown face could lean on a word, a phrase, and milk it for the most. Even his crazy coonass backwoods violent streak took on a certain charm when so well told: "I stabbed some people in my time, Hoss. Some for love, some for money, some just out of the plain meanness of my heart. And the funny thing about cuttin' is, once I start workin' that blade I can't stop. One man I left with his guts in so many little red ribbons they couldn't have put him back together with a spoon. But when I took that first slice through him, *mmmmMM!* I just kept on and kept on, couldn't nobody've stopped me with a steam shovel. It just goes to my head some way."

I cannot now, laying down Gitzy's rap in black type on a white page, call it lovable. But over the galley table, eye to eye, I was mesmerized. I thought Gitzy would make the perfect biographer for Charles Manson. So yes, I knew I was in the presence of living evil, but I felt lucky to be there. Here was a glittering vein of the stuff I love the best: writer's material. When I was not too drunk or too caught up in the tales, I ran back to my room from time to time to write it all down. I labeled my *Condor* notebook Sea Horror Dialogues.

I came to recognize a special category of Sea Horror Dialogues, the Gruesome Contest. These contests are common enough in the merchant marine, striking up at the galley table at suppertime, then floating out onto deck for the last sun of the day, growing more grisly and elaborate as passersby are drawn into the circle of contestants. Untrammeled by considerations of relevance or truth (the merchant marine I.D. card, called a Z-card, is also called a liar's license), sailors, rigrats, and dockies spin their tales until duties call them away and darkness falls.

Over beans and rice Mick set off one such contest mildly enough with an oratorical tour of an orange juice factory where he'd worked during the year the Coast Guard had suspended his license. He told how the concentrate is frozen in four-foot-by-three-foot slabs and stacked on pallets to sit until it's needed. "Summa that concertrate is six years old," Mick told us, as awed as he meant us to be by the marvels of modern technology.

SpaceMan picked up the conversational ball to lie that he'd once found a wood roach frozen in a can of lemonade concentrate.

"You shoulda sued the piss out of 'em," Kewpie vowed, manfully. "I sure woulda sued the piss out of 'em."

Then Kewpie himself went on to tell about a tarantula that jumped out at him from a box of store-bought bananas.

"Them thangs don't really bite anyhow," Gitzy put in, stealing Kewpie's mild thunder.

"I tell you about how a friend of mine's aunt put her poodle to get warm in her microwave?" SpaceMan lied again. "It exploded all over the place."

"Thass nothin'," Mick snapped. "I heard how one woman put her baby in her microwave to dry after she give it a bath. They were gonna lock her in jail, but she nailed a lawsuit on 'em bigger than shit."

Kewpie said he'd heard over the ship radio that a welder on our rig had got his hand crushed off in an accident that day.

"That ain't so bad," Gitzy said. "It was last year we was offloadin' pallets of chemicals and one-a them bags goin' up on the crane ripped open. There was this roustabout down on our deck and some-a that caustic soda? Spilled right square in his eye. He was screamin', goin' on about his eyes, his eyes, but hell, there wasn't nothin' we could do for him. Took him off the rig in a helicopter in the mornin'."

Mick formed a fist and shook it at the rig above us. "They didn't take him right off to the hospital? Night or day they woulda got me to a doctor or I woulda jumped up their butt."

"Sue the piss out of 'em," Kewpie chorused.

"Back in seventy-two," SpaceMan said, "I was working on a rig doing carpenter work in the galley. I was carpenter's helper then and the carpenter? He took and cut right through his finger with that circular saw he had. He didn't want to look at it himself so he stuck his hand up in my face and asked me, 'How bad is it?' I seen that old finger just hanging by a thread and told him it looked pretty bad. He just reached around behind him with his other hand, pulled that finger off, and threw it over by the fridgerfrators, said he guessed it wasn't any good to him that way."

"Thooooo," Kewpie whistled in appreciation.

Mick cut in again. "He was stoopit. Coulda saved that finger. Doctors now can sew 'em back on."

The pitch and pace of talk picked up then, leaping ungracefully from one sordid anecdote to the next. A parade of bloody specters made the scene: bare feet impaled on catfish spines, tongues foreshortened in confrontations with frozen metal, legs lost to mysterious and sudden cancers, flesh burned away in flames that engulfed a cratered-out oil rig.

I moved a few steps off, to the bare edge of earshot. The gore was getting to me.

Gruesome Contests are seldom concluded decisively anyway. There is no prize, no victor, only the game. This night of which I speak was a climactic exception. That master storyteller, our mate Gitzy, had moved outside the circle, too. The boys closed it up behind him, intent on their game. Gitzy paced the perimeter acting nonchalant, but he had one ear cocked, listening.

Then his eye fell to something on deck. A black beetle, one hundred and ten miles from home, was ticktacking its way along the edge of a deck board. Gitzy stooped and plucked the beetle up, much as one would nab a crawfish. Straightening, he snapped off its head. The men heard that sudden click and looked up in time to see Gitzy pop the bug into his mouth with a flourish.

Mick and SpaceMan laughed. I gasped. Kewpie cried out, "You ain't gonna eat that thing!"

In answer, Gitzy chewed once, twice, then extended his meaty tongue to show the broken remnants of his snack. He swallowed loudly and put on a farce of puzzlement.

"Whatsa mattah, Hoss? You don't eat bugs? Them beetles mighty good eatin', mighty good eatin'."

The winner.

19

Four or five days into the endless party and I stopped drinking. The supply of alcohol was running low, Mick and Gitzy had begun to guard it, and I was tired of headaching. I started back to work on deck a little, not a lot, just enough to keep my hand in. In the early morning hours, when I liked to catch the sunrise show, the men who had stayed up late around the galley table slept in. Karen and I had the boat to ourselves. She scrubbed down the galley, I hosed off the deck and painted until the sun was high and the boys were awake, ready for a card game. This became, after a while, our regular routine. We women, who shared a room, went to bed early to read and keep our diaries, and the boys stayed up to trade he-man tales and make a goat of Kewpie. Long into the night we heard them hooting and howling. I learned to sleep with two pillows over my head and one pillow under it. Karen tossed and turned, irritable.

On the night that I sat up late to make my notes on the Gruesome Contest, Karen had a bad case of menstrual cramps. I saw her toss and turn and toss again while Mick's voice boomed through our door. The joke: "You know what they do with the foreskins after they circumsise 'em off of little boys? They plant 'em in the ground and let 'em grow up to be big pricks. Har har har. Then they make 'em boat skippers. Har har har har."

Ten minutes later and the boys' laughter erupted into a roar. Karen swung out our door into the galley with murder in her eye. "How long is this going to go on?"

But her face dissolved into a smile in spite of herself. Mick was hunching on the floor wrapped in a yellow blanket. She asked what he was up to.

"Playin' caterpillar," he said. "You missed me doin' the pregnant gook woman."

Overheard in the aftermath: "We should have one of them movin' picture cameras. We could make our own *Saturday Night Live.*"

The next day we ran out of booze. One hundred and ten miles out in the Gulf of Mexico, fourteen days to go, and only a sticky deck of cards and Kewpie's hazing for diversion. Our television couldn't pull in even one clear signal. Things looked bleak.

But Mick had a plan. He would talk the rig's oil company man into letting the *Condor* take a one-day fishing trip. We would then, according to Mick, catch enough grouper, jackfish, and red snapper to trade for the booze Mick was sure somebody up on the rig was hoarding. An unlikely plot, but Mick brought off its first step. The company man, and we could almost hear him drooling for fish fry over the rig radio, gave us permission to go.

Funny thing about that, too. The *Condor* was the rig's permanent standby boat. Only when we ran out of fuel and water, and then only if another supply boat could stand by in our place, were we allowed to run to dock. It was part of the boat company's contract: the standby boat was to be on hand and at the ready twenty-four hours a day seven days a week in case of an ugly emergency on the rig; to fish burning rigrats out of the water if the rig cratered out, to evacuate the rig's crew if a hurricane came, to rescue any rigrat who plunged off the rig and into the Gulf. The oil company man gambled with his men's lives for a meal of red snapper. Mick gambled the Byrd Marine contract for a little sport. Both of them gambled that no disaster would strike, that no higher-up would hear of their chicanery. My own responsibility was clear: if the captain says we fish, we fish.

The fishing ground we headed for was what Mick called old iron, an oil-pumping platform that had been in place for more than ten years. Snapper, jack, and Gitz's favorite, shark, were known to congregate around old iron. We cranked up the diesels, lifted the buoy line, and headed out.

Karen, in an unusually sociable mood that day, joined me on the wheelhouse roof where we declared Ladies' Day and stripped off half our clothes to bake under the white sun while the *Condor* cut a twin white line through seas as blue-green and smooth as a hotel pool. With the wind cooling us as fast as the sun could heat us and lunch settling in our bellies, we exchanged girlish confidences. SpaceMan, I told her, had started snuggling up to me during the card games, calling me "Snookums." Karen admitted she was "terribly attracted" to Gitzy. She'd never thought "someone like him" (as if there might be more than one of Gitzy! Dear God!) could "get to" her.

I pictured that unlikely pair on land: Karen pressing wild flowers in her *I Ching*, Gitzy roaring from Morgan City low-life bar to bar on his hawg of a Harley. "And the beat goes on," was all I could manage to say about it.

"Gitzy kissed me last night," Karen told me when our laughter subsided. "I was standing at the stove and he just did it. Man, is he a good kisser! Then five minutes later I looked at him, and God! He looks like he stepped off a Ten Most Wanted notice in the post office . . ."

We laughed again and wondered to each other how two perfectly sensible earthwomen could lose their perspective, but good, once they went to sea.

"Still," Karen said, "when he kissed me my toes were shaking."

I'd been watching their little romance develop, the way Gitzy censored his meanmouth and his muggy faces when

Karen was around, the way he trembled with awe when she smiled her airy, benevolent smile at him. I'd caught him actually doing some work the morning before: taking out Karen's galley garbage, helping her reorganize half a ton of canned goods. He'd been calling her "Bootiful" of late. I decided I'd start bunking in the wheelhouse, to give their demented infatuation room to grow. That was a mistake. I made plenty of mistakes on that hitch.

Three hundred and sixty feet of water at the platform where we tied up to fish. Clear water, too, as Mick demonstrated when he dropped one of our heavy china coffee cups over the side. I watched its white blur all the way to the bottom. Fish ganged around it when it settled. "Hot dawg!" Mick shouted. "See them snappers? Must be ten thousand of 'em!" Everyone ran for tackle and bait.

Gitzy's tackle box was downright scary. He had an inventory of hooks as shiny as tailors' shears, as big as stevedores' grappling tools. "Them's for my sharks," he said, rummaging beneath them for the smaller snapper hooks.

While he put his reel together he instructed me in the construction of a simple hand line: four hundred feet of thin cotton line with a snapper hook tied on and nuts and bolts for sinkers. No bobber at all because red snappers are bottom feeders and they won't hook themselves. When I felt a tug on my line I was to tug back, sharply.

Feel something at the other end of three hundred and sixty feet of twine at the bottom of an ocean? I doubted it. But I speared a shrimp for bait and tossed my line over the side. In a minute I felt the sinker hit bottom, and that was reassuring. Gitzy told me to haul in three or so inches of line, and when I did that, *zing!* A bite. I jerked back on the line, purely a reflex action, and hauled the line back to the surface. There he was: my first fish, a gleaming red snapper, so beautiful, so alive, that if the boys hadn't seen him I'd have set him loose again. But this was our lucky fish, first catch of the day. Mick helped me loose him from my hook and left him to flop out his life on our superheated steel deck. I speared another shrimp and lowered my line again. Again *zing!* Another snapper.

The boys glared at me. Still fussing with their rods and reels, they hadn't even dropped their lines into the water when I hauled up my third snapper.

"It's so easy!" I said.

"Grrrr," they growled.

Mick's heavy-duty line broke when he got his first bite and there was a *whirr* of speculation among the boys. Must be something god-awful big down there. "Grouper," SpaceMan guessed. "Shark," Gitzy thought.

By this time Karen had fashioned her own handline and hauled up a couple of snappers. Girls 7, Boys 0. Seven big beauties, all still slapping themselves to death on the deck. I didn't like that part.

"Fish don't feel no pain," SpaceMan promised.

My fish eyed me, telegraphing quite another message. Karen, as softhearted as me but more experienced at fishing, fetched a ten-pound sledge from the boat's paint locker. She conked her fish and turned the sledge over to me. I was conking mine squeamishly, shivering at the sight of their eyes popping loose and flying over the deck, when the boys saw what I was up to.

"OOOGH!" SpaceMan shuddered.

"How come you doin' *that!*" Mick howled.

"I'm putting them out of their misery," I said, conking the last one twice.

"Bloodthirsty for a girl, ain't she?" Gitzy leered.

But the fish had stopped slapping. I thought, but couldn't be sure, that I had done the humane thing. When Karen hauled up her next snapper I bopped it for her, more accurately this time.

Mick ripped the sledge out of my hand and sailed it over the side. "I told you to cut that out," he hissed into my face.

"All right, all right," I backed off. "Sorry." But it was odd, how those rough-talking sailors blanched at seeing a woman kill a fish. Odd. The men stayed uneasy for another while. Karen and I, and after a while SpaceMan, too, kept hauling up the snapper. I worried about the way Mick wouldn't quite look at me. I'd offended something deep in him, and couldn't understand it.

But when I turned to throw another snapper on the deck I saw that Mick was waiting for it, in something like good humor again. "You women think you're tough. Watch this here."

Mick produced his wicked-looking fishing knife, stripped the scales off my snapper while it struggled in his hands, then took a bite out of it. Mick was, remember, toothless. That didn't stop him. *"Graaa!"* he roared.

I snatched up my camera and got a picture of him doing it. No one back north would ever believe this, I knew. Man bites fish.

I went back to my fishing. But either we'd cleaned out the available snapper or frightened them off with our depredations. All I could hook were what the men called trashfish: a shimmery blue and yellow triggerfish ("Tho' 'im back, don't even make bait," Gitzy said. "And watch the mouff on 'im, he bite."), a slew of baby groupers who rolled their rabbity eyes at me expecting the worst, and a long black snaky remora, the tiger shark's pilot fish. Gitzy greeted this last with a whoop of delight. "I knew them tigers was out there. Now I'm gonna get me one. Come on, Hoss," he said to me. "I'm gonna teach you how to get a shark."

The hooks I'd seen before, the ones that looked like you could hang a whole hog from them, Gitzy tied with heavy-gauge wire to a half-inch nylon line from the ship's stores. Near the other end of the line, for bobbers, Gitzy tied three empty one-gallon milk containers, "for drag," he said. When the shark took his dive, these hefty bobbers would slow him, eventually wear him out.

Then, over Karen's protests, Gitzy impaled an eight-pound bleeding beef roast on the hooks and cast the whole nasty mess over the side. He tied the loose end of the line low down on one of our eighteen-inch steel bitts and lay down beside it to wait.

"Won't be long and I'll have me a big old shark," he said.

While the rest of us cleaned our snapper and washed up for dinner, Gitzy stayed on deck, eyeing his line. Karen took his fish fry out to him later, when the sun went down. I moved my bed on the wheelhouse bench. This looked to be the True Lovers' night for romance.

"He sure does have a lot of energy," was all that the wan-looking Karen would say about it the next morning. I went out on deck to my usual rinsing and painting and saw, at the end of Gitzy's shark line, a shark. I don't know why I was so surprised.

When I shook Gitzy awake I said, "You got one!"

He knew just what I meant.

Together we wrestled the thing aboard. "Got the ugly sucker, didn't I?" Gitz was in a frenzy, sweating with the effort, the joy.

The monster he had caught was a hammerhead shark. I'd seen pictures of them in books, but nothing could have prepared me for this reality. It was just a youngster, maybe seven feet long, but sleek as steel and thrashing like spring steel, too. Gitzy said it was worn out, done in. Its eyes said something else. Two thick fleshy posts protruded at right angles from its curiously narrow head. At the ends of the posts, eyes big as fists, and although alive, still dead as the eyes of a mad-dog killer subdued at the end of a murder spree by a tranquilizer gun. Eyes that put a chill on me. They did not, like the snappers', seem to beg for mercy. The shark thrashed wildly on the deck, and still the eyes seemed dead. The hooks on which this beast was caught protruded, one from its soft rippling gills, the other through its lower jaw. Brrrr.

When I snapped Gitzy's picture with his prize, he posed like Teddy Roosevelt with the horns of a wildebeest between his hands. Grinned like Roosevelt, too, a small square grin.

He planned to boil the shark's eyes until they hardened into stone, but Karen squelched that idea. Not on her stove he wouldn't. He had to content himself with ripping the jaws out of the still-living shark. I gagged and ran back into the galley.

"Karen, you won't believe what he's doing."

But when I told her, she only shook her head. "Sailor's ancient enemy," she said, unsurprised.

We sailed back to the rig late that day, all of us silent for once, logy as a family of campers coming home from Labor Day weekend. The postsunset sky was a cumulus nimbus brocade in salmon pink, deep velvet violet with a streak of surprising neon green beneath one flickering star. Venus, Karen said.

Captain Billy had taught me how to recognize cumulus nimbus: by the animals on top. That night's menagerie included a scorpion curled to sting and a flying fox with an eye on his pursuers, a determined spermatazoa and a sea turtle snapping her jaws just sky-inches from the fox's fan of tail.

The seas were ruffling up for a storm show and I marveled again that I was being paid, forty-five dollars a day now, to live aboard that boat, at peace.

The next day, when Kewpie left for home, the crew of the *Condor* turned on me.

20

*K*ewpie's replacement, Danjones, called by his whole name, was an old regular with the *Condor* crew and a special friend of SpaceMan's. I was glad to see him come and Kewpie go. But Danjones did what Kewpie had done first thing aboard: shoved me away from the rig line I was loosing from our bitt. "Stand back, honey," he said.

I did what I had done with Kewpie: shoved him back and cursed him out loud. Our struggle lasted no more than ten seconds.

But when I turned back to finish the job on the line, I'd lost my slack. The *Condor* was drifting away from the rig, putting three hundred tons of pull on the line. Even a tough six-inch-diameter rope will snap under that kind of strain. This one was groaning, getting ready to pop. Nothing unusual, it happens all the time, but when it does, the deckhands standing near had better back off quick or hit the deck flat because, as Gitzy put it, when stiffline pops, "it turns to glass on the broke end and whups around—put your eye out quicker'n you can say beans and rice." The hands get plenty of warning: that snarl and groan, then *whup-whuuuupp-hhup-pup-pup-Pop-POPPA!*

I backed off fast, and nearly collided with Danjones. Veering around him, my bare feet flew out from under me, my bottom hit the deck, and the back of my head smacked into a stanchion. I got back to my feet and tended the line, which was going slack again thanks to Mick's giving the diesels some juice in reverse. The fall was no big deal, to me. Lines pop, deckhands fall; it happens. I finished loosing the line, signaled the crane to haul it away, and went about my business. The accident was minor, and nobody's fault. Half an hour later I would have forgotten the whole thing.

But this very minor accident had some special mojo for the men. I overheard the whole crew discussing it, worldly wisely. Gitzy saying I work too hard. Mick insisting I was fooling myself; no woman ever made is strong enough to handle a deck job. SpaceMan saying I'd been getting above myself, trying too much too soon. Danjones' only comment was, "Yankees!", delivered with disgust. Even Karen chimed in: "I'm just afraid she's going to hurt herself out there throwing those heavy lines and working under that big old crane."

Pinheads.

No sailor, even a cook, escapes an occasional pratfall. Why was mine suddenly such a big deal? No one in the crew had made much of my being a woman in a traditionally male job until that moment. I worried that thought around and then blew it off. If I didn't make an issue of it, they'd find something better to talk about. I went back on deck and put in a couple of hours of painting, taking up where I'd left off on the diesel stacks.

That afternoon the oil company man called us on the radio saying he had a taste for red snapper again. We cranked up the diesels and headed out.

Mick knew of another piece of old iron where the fishing was good. But it had been so long abandoned he worried we wouldn't be able to tie up to it. Its only bitts were high up on the iron. On our way there I practiced my lassoing again. I'd lasso that bitt the first time, just to prove them wrongheaded about women as deckhands.

And I did it, too, on my first try, lassoed a bitt so high on the platform we couldn't have reached it with our eighteen-foot gaff hook. Feeling smug, I reported back to Mick. I thought he'd be proud of me.

"You and I are getting too old to do that kind of stuff," was all he said. Implying that maybe I'd accomplished this near-miracle today, but tomorrow?

I was pissed. While they fished the next day away, I painted. Paint, paint, paint, spite, spite, spite. No transcendence this time.

That afternoon the rain came. When the galley conversation turned to whores, wives, and how a woman ought to stay behind the stove where she belonged, barefoot and pregnant, I went to the wheelhouse and looked out over the water. A circle of rain-shot clarity extended a few hundred feet in all directions; beyond it, blue-white nothingness. Sailors were always saying that this aloneness on the great waters is what kept them coming back to the boats, this sense of being the only one of your kind for as far as the eye can see. Their thrill, not mine. Being alone like that made me itch all over. Just the day before, my crew had been my friends.

Over supper, Danjones mentioned there'd been a rape on the beach in his hometown, Biloxi.

"Rape?" Mick boomed. "Rape? I never could understand all the fuss about rape. Me, I'd rather fight than fuck."

"Yeah," Gitzy mourned. "But a good fight these days is harrrrd to find."

I'd begun to suspect, after six months in Dixie, that its male-to-male random violence was a major cultural pattern, no accident at all. Now here was Mick saying he'd rather fight than fuck—a major dislocation in the hierarchy of needs. My ears pricked up. I asked questions. What made a fight a good fight? Was that the same as a fair fight?

Mick said, "Hay-ell no! Ain't no fair fights left anyhow. Whole hog, root or die, thass a good fight. 'Nother thing about a good fight: it can start up over any damn thing, but it never gets over with.

"You come back next week. He calls his uncle. You call your friends. You hear what bar he's drinkin' at, go over there early and dive in, the whole bunch of you. And I don't stop at no fists, no. Not me. I go home for my forty-five, come back and cover the whole barroom with one barrel.

"Now a fair fight, I haven't seen one of them in so long I can't remember if I ever did see one. You, Gitzy?"

"Not me neither, Hoss."

"Me, I mash my foot down in his balls straight off, and keep on a-kickin'. *Kyeeow!*" Mick howled with his round eyes rolling. "Bit the end off a fellah's nose once, sure enough did. Had a holt of him like this"—Mick demonstrated on SpaceMan—"and all I had left to hurt him with was my teeth. *Graaaah! Shnip! Ptooo!!* Har har har. I had me some teeth then, good teeth. Jumped right up off 'im and took off outa there before his buddies could stomp my ass.

"I'm not one of them it matters whether it's man or woman. I'd soon deck a woman as a man. *Kyeeeow! Thomp!* Knock her right out and keep comin'.

"*Theeoww! Kyeeoww!* Take you"—Mick turned to me—"and that night you took into me about the cards." He referred to the first night we played cards together. I'd squawked when Mick arbitrarily awarded himself two hundred points for being captain.

" 'Member when we sat there head to head and you said what you did and I th'ew down my pen? That there was the sign—you musta knowed it, too, 'cause you took off to your room then—that was the sign I was gonna knock you plum into the sink and mash your teeth down your throat right after.

"See that old fire axe over there? I woulda cut off your arm clean through the bone if you had a-kept a-comin'. It's just a good thing you lit out when you did."

Gitzy nodded vigorously, seriously, remembering. "I knew it. I seen it comin' when he throwed down that pen and you saw what I did then, didn't you, Lucy? I got back out of the way."

Mick's eyes had not left mine. "Woman speak up to me like a man, I'm gonna treat her like a man and just stomp the livin' piss out of her."

I tried to hold my face in place for this scary showdown but came the queasy-weasy sensation of my heart's sinking, the involuntary drop of my jaw, the sting of my unbelieving eyes. The sound of no pin dropping. Mick's threat was real: a beating with dismemberment thrown in if I dared "speak up to him like a man."

"I sure would have been *surprised,*" I managed to answer.

Around the table: "Har har har."

But this was no joke.

Every day I could feel my position in the crew deteriorating. The boys, which is to say SpaceMan and Danjones, took over what little there was of what they called "the real work" on the boat: the tying and untying of the rig's lines. The remaining deck work, hosing, painting, scrubbing, they left to me. SpaceMan stopped joining in with the indoor sanitary job. "That's women's work anyhow." Mick condoned him.

Whenever we went under the rig I showed up at my post at the stern to help with the lines. The boys shoved me aside, and harder than they needed to. I complained to Mick. He ignored me. At the brink of frustrated tears, I went down to what had once been my room to talk with Karen. She was no more sympathetic.

"Is that what you want? To get hurt? Because otherwise I'd just let them tend to the lines if I were you. You have enough to do with the painting."

I closed the door behind me, shut myself in with her, and cried, shamelessly. "I don't think you understand," I slobbered. "It's my *job.* If they keep me from doing it, they can tell the office, truthfully, that I'm not doing my *job.* Then I'll have to go back to *cooking,*" I bawled.

Karen slammed out of the room. That was our last talk aboard the *Condor.* That was, now that I think of it, my last

real talk with anyone aboard the *Condor.* Seven more days
to crew change.

I retreated. Backed out of the card games, backed away
from the idle, vicious chatter at the galley table, withdrew
from the fishing, fled from meals and mealtime banter where
the men pressed me to the wall with what had become the only
subject in town: personal, institutional, and recreational violence. Danjones telling us the way to go about stabbing an
alligator so it can't bite you back. SpaceMan boasting of his
motorcycle pile-up, parting his hair with his fingers time and
again to display his knotty violet scar. Mick replaying an old
bar battle, using the beam of the engine room flashlight to
illuminate the cavern of his toothless mouth where an apparently even uglier scar testified to his badly broken palate. I
didn't look, but the boys did. Har har har. When they laughed,
they aimed their laughter at me. I was the new bottom dog.
Har har har.

Tying up to a rig is difficult, is dangerous, is part of everyday deckhand routine. The rig lowers its two lines, one at a
time, to the boat's back deck; the deckhands must secure the
lines to the boat's bitts with four or more figure-eight wraps,
and with the correct amount of slack line between boat and
rig.

The first step of this dangerous dance is to get some slack
line onto the boat and start looping what they call the standing part of the line over the bitt as the captain holds the boat
in position.

Deckhand Number One does a quick loop of line over bitt,
looks back at the captain on the stern controls. The captain
signals whether he wants more or less slack in the line.

Deckhand Number Two looses the crane's hook from the
line and stands by to "give slack" to Deckhand Number One.

The boat engineer usually stands near in case the deckhands get into trouble.

The line is four to six inches in diameter, usually wet for the first twenty feet or more and heavy as a Victorian back bar.

It's just crucial never to let a long slack of rope hang in the water. That's how a boat gets a rope in the wheel. Rope in the wheel is the Gulf Coast signifier of poor seamanship, the bane of captaincy. So the tying-up process must go swiftly and surely.

For one deckhand alone to attempt tying up to a rig would be dangerous folly and as unlikely a sight as you'd see on the Gulf.

I'd been a full-fledged member of the tying-up team until Danjones arrived. Now, crowded out, I made a last and loud complaint to Mick. I wanted to continue doing my rightful part.

Mick's reply was a surly, "Well, if you want to be a man, get out there and be a man. Damned if I care."

Next time we came under the rig I went to my place. Mick, at the stern controls, maneuvered the *Condor* into position. No one shouldered me aside. All was as it should be. The crane operator ten stories overhead lowered the first line. It was just an arm's-length away when I looked around to see if SpaceMan was in position to give me slack. But SpaceMan, and Danjones, too, were shuffling their feet and staring out to sea from the far side of the deck.

Oh. I get it. Another No Girlz Allowed plot to put me down for good and all. They knew I wouldn't be able to do it alone; I'd have to cry for help.

I didn't have a lot of time to think it over. The rope was snaking down at me, ominous as a war party overlooking a lone cavalry troop. I grabbed it up and went on alone. Damned if I'd beg those bad boys for anything.

I got the line looped once, dashed to the stern to flip my slack line to the outside (oomph), rushed back to the bitt and pulled up my own slack, working both sides of the bitt tediously. Slow-motion Wonder Woman. You remember the mother who lifted the Volkswagen from atop her crushed and dying son? That was the stuff I was made of for those year-long moments: one-hundred-proof adrenal plasma. Fifty pounds of wet rope at a time I pulled up, stepped over, wound into figure-eight loops on the high bitt.

I did do it, too. Alone. The one time I had a split second to check on the whereabouts of the rest of the team, I saw them composing an elaborate Concerto for Whistle into Upwind. So I went the last mile, tightened the wrap, and it didn't even look bad. Nor did I lose an arm or a finger, two potent possibilities.

That done, I hefted slack to SpaceMan while he tied off the line on the opposite bitt, doing my best to keep my ragged breathing inaudible. That done too, I stood back and waited for acknowledgment. The bet had been tacit: she can't do it. But I had. Still, SpaceMan didn't say a word. If anything, he looked disgusted. Disgusted?

I huffed off to the galley to wash up.

Gitzy was crouching at the galley table, slap-shuffling a deck of cards. After a minute he looked up and stuck his lip out at me. "I hope you learnt out of doin' that."

"Learned what?"

"That you can't do it alone."

"But, Gitzy, I did. Do it. Alone."

"Yeah, maybe you did it this time, but it was a damn fool thing to do and it could get you in a shitload of trouble, Hoss."

Mick shuffled down the steel stairs and fell heavily onto the galley bench. "I'll say one thing for you, Wucy. You got the determination. Maybe you ain't got nothin' else, but you got the determination."

The remainder of that hitch I spent on deck, finishing off my paint job. The water was silent, almost still, only occasionally breaking over the deck in even, unruffled wavelets. When I plowed back into my *sesshin* I noticed that dreams I couldn't recall on waking came back in full color and visual detail. My dream guide had returned one night to teach me how to survive the taunts and threats of this crew. He pointed out Everything I Want and Need, a blur of island, presumably land itself, just at the horizon of my dream. I felt a chill of fear at his innocence, at his trust in the things of land.

My dream guide came once more before I debarked, with a dreamy slide show this time. The only way to get through to these people, he transmitted to me, was to restructure their chromosomes microsurgically. I was tickled by that dream. Perhaps the extreme heat and barometric pressure of the Gulf Coast warped the men's little XY's into XYY's and even XYYY's. (Certain contemporary scientists claimed to have found such double and triple Y chromosomes in blood samples taken from sex criminals, hard cons, and killers.) Naturally when my little XX feet stepped aboard, the boys formed up to wave their Y's at me. This was only one of many peabrained theories I'd come up with to explain the men's behavior, all my theories stopgaps against the brutal realization that this dream of mine might never work.

On the last night of my twenty-one with Mick and his crew I climbed to the wheelhouse roof to watch the wicked and definite morning glory purple of the after-sunset. In its foreground, parlor-pink whuffs of smoky light were scattered by sabers of another pink yet. The boats and rigs I'd missed seeing with a naked eye appeared as beads of color, flicking on in sequence like the lights in department store windows. I spied Danjones idling on the barely rolling deck, venting six hundred gallons of raw sewage over the side. Strictly illegal, but less trouble than hooking our sewage pump to the dock's tank when we docked next morning. That's how it goes, on the Gulf.

Crew change was only hours away. By that time I would have traded mine for a bed full of bugs and burrs. Bad boys, looking to put me in my place, had found me out of reach and ready with an evil eye during the last days, so they turned on Karen, stepping up their lethal true-tales. Even Gitzy, formerly Karen's protector, had begun to aim his ugly talk at her: "They know I done it. Everybody knows I shot that nigger hitchhiker in the face but they can't find the Magnum I done it with." The sounds of Karen's weeping rang through the boat's ventilators.

When full darkness came, I climbed back down from the roof to my private nest in the wheelhouse. I'd saved this best moment for last; I was going to look at myself.

Mirrors always have spooked me. When I was an unhappy teenager I used to be able to count thirty-four physical defects in the mirror over my dresser. Later, during an acid trip in 1967, I stared into a medicine-cabinet mirror and watched my face melt into puddles of raw gore. There are maybe twenty photographs of me in existence; cameras I like even less than mirrors. This is a negative vanity. In my four weeks aboard the *Condor*, my body had changed. I knew I'd dropped some weight; I was having to tie up the waist of my cutoffs with fishing line. My ankles and arms were red-brown, skinny, glowing with—what could this be?—health? What I'd been able to see of my hair was shockingly white. But since the only mirror aboard was in Mick's stateroom, off limits, I had only those few clues to my physical transformation.

I put "Thunder Road" into the tape deck and adjusted the wheelhouse lights to make a reflection in the big square windows. With the black night sky behind them, I had mirror enough. Gads.

I'd shrunk from a size fourteen to what must be a ten. My striped lavender-and-white T-shirt fairly glowed in the dark next to my Coppertone billboard tan. The sun had bleached my hair to Johnny Winter white. I rolled up my pants and sleeves: Jesus! Muscles! Not unattractive, either. Nicely articulated, clean, flexible. I claimed to be thirty-six years old but the mirror woman was some cross between fifteen and sixty.

I stepped closer to the reflection. Every pinch of what I'd always called baby fat had burned away from the bones of my face, the sinews of my neck. Nice. But the mirror woman's black eyes, volcano eyes, warned me off. And what had once been laugh lines, bisected by dimples, were lines of menace now. Here was a woman who could, it was plain on her face, wring a neck on a moment's notice. Those lines. Those eyes.

I readjusted the lights and sat back down on the wheelhouse bench out of reach of the mirror woman. Here were my notebooks, my friends, my pen mightier than Gitzy's evil tales, Mick's threats. I scribbled an account of my last day aboard the *Condor.*

Mick appeared, and after an uneasy moment's silence, asked me what I was writing in my book. I didn't even look up. After a while he wandered off again, calling over his shoulder as he descended the steel stairwell, "Why can't you just be a woman?"

I ignored that, too.

The galley-to-wheelhouse phone rang. A call from Space-Man. In its entirety:

ME:	Yow?
SPACEMAN:	It's me.
ME:	And?
SPACEMAN:	You drunk or something?
ME:	Nope. Just about to smoke my last jay.
SPACEMAN:	Well, you wanted to talk to me?
ME:	Who said?
SPACEMAN:	Gitzy said.
ME:	Gitzy's just messing with your head.
SPACEMAN:	(pauses, breathes) Well. Do you? Anyway?
ME:	(pauses, prepares) Talk to you? No. I wouldn't mind dropping you over the side, but no, I don't want to talk to you.
SPACEMAN:	Oh. OK.

An hour later Mick was back, hanging over the rail on the deck below me, perched on my freshly painted bitt, staring out into the dark Gulf. He said, loudly, so I could hear, that he was sad.

I didn't say anything. I had my own feelings to deal with. I had just that moment caught myself composing a goodbye speech: "Gentlemen, I leave this boat hurt, puzzled, and none the wiser . . ."

Danjones, finished pumping his sewage now, visited next. He said something friendly, something normal. I looked up into his clean-cut American boy face, flashed my volcano eyes. The eyes said it all: move off.

He moved.

There was a hand-drawn cartoon taped to the *Condor*'s wheelhouse wall, a cartoon that got me on my nerves. In it a pair of naked toddlers, one a boy, the other a girl, faced a badly drawn cartoon bathtub or toilet. In a bubble over their heads the cartoon boy was saying, "No, you can't play with mine. You already broke yours off." This would have been the moment to rip that thing off the wall and toss it to the wind. I wish I had.

Mick, still hanging on the bitt, was weeping loudly before long. Finally he approached me, dangled himself over me and my notebook. He was hard to ignore, but I went on writing. What was there to say, to him?

He said to me, "Don't you want to know *why* I'm sad?"

I looked up, and waited for him to say it.

"I'm sad 'cause I'll miss my wittle Wucy."

I never could manage to sleep on the eve of a crew change. Crew change is almost a death, the way the empty sea can be. The tribe, however ill mated, has crewed, broken bread, collaborated, bullshitted together for so long that each of us, no matter how carefully separated, has formed a web of attachments with the others.

New characters would come to take our old places, robbing us of something that had no name, but it was real enough, to us. A new acculturation would commence, a fresh social order peck itself out. A new man, someday maybe a new woman, would take over my spot at the bottom of the heap. My dreams of finding a home at sea were going sour. But at least I was getting away from these bad boys.

21

*B*ack on the dock at Intracoastal City, I believed I had just enough time to visit Guste on the *Pride* before the Byrd Marine carryall arrived to take me home. But Guste was on home leave and I met the alternate captain who held his place that week.

"I heard about you," he said, leaning on the "you."

"Oh yeah?"

"There's a name on you around here, I heard it just the other day. They call you The Morphadite."

"Really? I look like a morphadite to you?" Pulling the lids over my volcano eyes, I allowed him to inspect me.

"You never can tell," he answered, finally. "But if you was my woman, I know what I'd do. I'd keep you barefoot and pregnant, the way women was made to be."

Barefoot and pregnant. I kept hearing it. They kept saying it, as if it were a sentence they could impose on me if I stepped out of line.

I had the opportunity to study, during my onshore weeks, a woman who lived such a life: Jolene, wife to Cupp.

When the Cupps moved into their brand new trailer, which they called their mo-bile home, they abandoned their humble tin-roofed four-room shack, and persuaded their landlord to rent it to me. Their trailer, and my shack, stood at the intersection of the old Spanish trail and the cattle-driving route from East Texas to New Orleans. That's the road the Cupps and I took to town and back when there was shopping to be done. The Cupps' youngest child, a son, Aron, sat half on my lap one night as Cupp drove us all home from one of our shopping trips.

"Lookit them wales, boy," Cupp teased his toddler son. "Just lookit them wales."

Aron cranked his innocent head around, peering out into the thrill of darkness beyond our windshield for a glimpse of the leviathans.

"Whut wales?" Jolene piped. "Whut wales, honey?"

"The way-ells they appear to be diggin' here in this damn road."

"Oh, them wales! Here comes another one along!"

Jolene is quick enough to pick up on any joke. And although two of her children, Aron and Treasa, will spend their lives with misspelled names, Jolene is only uneducated, not dumb.

She was a dirt-poor, half-orphaned thirteen-year-old north Alabamian child when Cupp, a sturdy twenty-three-year-old mechanic, married her. Jolene has known no other man in the seventeen years since, no other life beyond the four walls of her home in the swampy places of the South. Barefoot and often pregnant, mother of five, thirty years old.

Jolene navigates successfully enough in bank, supermarket, and post office. What's more, she can twist a log out of a brush pile and take a bead on a snake with the best of them. I've seen her do it. But the world beyond Lafourche Parish's public places comes to Jolene on the hearsay line, through *The National Enquirer*, daytime television serials, *The Thibodaux Comet*, and Cupp, her supreme authority.

Cupp told me, for instance, that when election time comes around he won't allow Jolene to vote unless her choice duplicates his. "Otherwise, she'd be like cancellating my vote."

Jolene nodded ferociously when he said that.

But with Cupp home from the water only one week out of three or four, Jolene manufactures her own bizarre certainties. The Commonists, she says, poison our local drinking water with That Blue Stuff, and blow up the oil rigs offshore. They run the Phone Company too, she suspects, and the Board of Education. Once she called her children's teachers to complain that three R's are enough for any child: "A kid in fourth grade's got no business studying science. It's jest one more excuse for them to get nasty."

"Make me President of America for one year and I'd have the whole mess straightened out," Jolene says. "You laugh? Jest watch me!"

Murderers run a close second to Commonists for daily nuisance. And since every murderer is a stranger, every stranger is, *ipso facto*, a murderer. I had all too few visitors to my little house, but Jolene was outraged at the traffic. "How come you want people to know where you *live!*"

Jolene herself is safely barricaded behind a wall of frustration and noisy children. She told me, proudly, that she cleans out her cabinets six days a week, rinses out her family's empty pop bottles twice, with Lysol, and puts five loads of laundry through her bronze-tone washer set every afternoon. I could hear her through the foil-thin trailer walls, screaming at her screaming children. "You gotta stop eating jest whenever you're bored! I won't have it!" I could hear, too, the reedy strains of her favorite TV serials' theme music. But Jolene told me that *Ryan's Hope* "is turning dirty. I keep half a ear closed when it's on. I don't like people talking dirty about filthy things. Nasty-nasty!"

Filthy things are all those things under your clothes. I remember believing, when I first met Jolene, that she and Cupp were sexually close. Seeing them lounge together on their two-hundred-dollar discount store sofa with his arm looped behind her, her hand resting on his thigh, I found it possible to imagine them locked in sweaty coitus. Even Cupp, with his shock of coarse white hair, his pinched Appalachian face, his lanky body as white as a mushroom from working belowground in the engine room, even Cupp looked a little sexy cuddling with his wife. Jolene, normally a dry personality, seemed moist then. A soft haze spread over her high bronze hatchet cheekbones, and her sprayed and back-combed wings of Clairol gold hair came loose, just a bit, when she leaned over to squeeze his knee.

But later, when we became better acquainted, I was able to infer the sad truth of their love life. A black woman had let her in on the big secret, Jolene told me, the reason why men would rather bed black women than white. "You can prove it yourself," Jolene said. "When you were married, did you ever really feel his, you know, *thing* when it was, you know, inside down there? Well, black women *do*. They're built that way. Different."

These were the kinds of worldly pronouncements Jolene brought to me on midafternoons, during weeks when Cupp was gone to sea. She'd come knocking on my door, saying every time, like a little ritual, that she had heard me "typing too hard." I needed a break from "all that typing," she said.

"I'm a writer myself," Jolene told me the first time she caught me at it. She produced from her wallet a clipping so worn from folding and unfolding that it looked like a tattered paper doily. It was a letter she had once written to the editor of *The Thibodaux Comet*. In it Jolene had strung her disconnected ravings like mismatched beads on an outraged whine at the way things are going nowadays. She sure told the editor.

She sure told everybody, and lurked behind every corner of conversation to tell them all again *in absentia*, raving her ancient psychodramas to life in the discomfort of my own home. Too often she drafted me for the antagonists' well-worn roles. (They don't call it acting out for nothing.)

She once pressed me into playing the part of the hard-line Baptist grandmother who'd "hopped all over" Jolene at the age of ten for wearing the red dress of a temptress. Then I was cast as the doctor who wrote on Aron's chart that "mother seems unconcerned." She blasted his ears off, I promise you.

Jolene has never held a job, or engaged in a social life. And although she knows just about everybody on Sixty Arpents Road, she has never had what I would call a friend. She hasn't even had a breakdown, although she's been hospitalized twice on the verge of one. Once when I was unable to sleep, troubled by my ever more frantic missing-the-boat dreams, Jolene gave me one of her "nerve pills" and I was comatose for fifteen hours. Haldol is what it was, the drug of choice for manic depression. Jolene pops four or five of those candy-colored life flatteners every day and still has enough energy left to rave the hours away.

Cupp alluded to this state of his wife's psyche only once. "Y'know," he mused, "Jolene just hasn't been the same since she got that misterhectomary."

It's easy to make a joke of Jolene's life, but the truth of it is urgent, bitter. She told me she wants to know what she's *for*, now that her supply of babies has dried up. "My womb jest popped there when Aron was born. It sounded like a shotgun went off inside me, I swear it did. *Bam!* Like that! When I got home from the hospital my womb fell out. I pushed it back in and called the doctor, but it fell out again and turned black and the doctors had to cut it out. What am I supposed to do now?"

I watched Jolene contemplate stealing an infant she'd been asked to care for while its mother was in the hospital. "She's already got fourteen of them, not that she kin even keep them clean let alone take care of this pretty new boy baby. I don't think she'd even notice if I kep' him."

But while Jolene dreamed of stealing the child, she also schemed to escape. She confided to me that she was putting money in a private bank account, unbeknownst to Cupp. "How far away do you think a person could get on a hunderd dollars?" she asked me once, as if apropos of nothing.

Later that week she asked me again, only this time she'd made a bank deposit and had one hundred and forty-five dollars on hand. How far would that amount take her? I couldn't begin to answer such a question.

But that was all right, too, because just a few days later Jolene called on me with the big news: a woman who'd had a hysterectomy had conceived a child. A white woman, too, Jolene vowed. The fetus had attached itself to a bowel, but it grew normally and came out all right, according to Jolene.

"It was in *The National Enquirer*, so it's really true. And I don't know if I should be happy or scared or what. What am I gonna do with a new baby, with the way things are going nowadays?"

22

The week I'd planned to spend at home turned into two weeks, then three. I'd asked Byrd Marine's personnel people to assign me to the first deck job aboard any of the company's twenty supply vessels, any but Mick's *Condor* or Billy's *Sandpiper*. But week piled on week and the closest I came to the boats was in my troubled dreams. I paced my house alone, did a little writing, and spent altogether too much time with Jolene. Every few days I called Byrd Marine, in case a deckhand failed to show up for his crew change. "We don't have a slot for you yet," the Byrd man said.

At the end of the month I hitchhiked down to Morgan City to see the Byrd man in person. Two young men in the company's lobby were scribbling their names on hiring forms. They'd been hired as beginner deckhands and were going to work that same day. Not on the *Condor* or the *Sandpiper*, either. When finally I was admitted to the personnel man's office, I had a hard edge in my voice.

"I wonder why you're hiring two new deckhands when you don't have work for me."

"Well, Lucy, we've been trying like hell to find just the right spot for you. Not every captain will take a woman deckhand, so it's a problem of *placement.*"

I told him I was packed and ready, that I'd be camping out

in the lobby until the next *placement* opened up. I had an assignment the next morning. It didn't last long. The captain slept across the threshold of my bunk room every night. To protect me, he said.

From that captain to the next, from one boat to the other, from the end of June to the middle of September, I was a moving target, a lonely vagrant on the Gulf.

There was the captain who would not allow me to work on deck. I was expected to clean the interior of the boat, and I could go on deck to do the laundry, but I was not allowed to "mess wid dem lines."

There was the captain who crossed himself when he saw me step aboard. He radioed the office to have me put off again.

There was the captain who sent the (male) cook out on deck to do my job, requiring of me only a plate of home-made biscuits for every meal.

Then there was the captain who never showed me his face. His wife had threatened to leave him if he allowed a woman to board his boat, so he hid out in his stateroom until crew change, when I was set ashore again.

Once I was assigned aboard a boat where I was ostracized, and that by total strangers. It was nothing personal, the engineer whispered to me when no one was looking. It was just that I was a woman, and women did not belong on boats.

That was a painful week. The men excluded me from the work, from the social gatherings such as they were, from the wheelhouse, from the engine room. As if I'd misbehaved and been sent to my room until I could come out, apologize, and somehow transform myself into a male.

No sooner would I step out on deck to chip and paint than the captain would tell me to "get back in the house where you belong." He was the captain. I couldn't ignore his order or even argue with him. I bottled up my anger, got sick with it, felt confused and soft and wrong. I remember vowing to my notebooks one afternoon that the men could go ahead and do their worst but I would stick it out. Sooner or later they'd tire of their game and accept me. It had to happen.

Our dock for that evening was the Baroid mud yard where we were taking on dry mud for delivery to Penrod 53, an offshore rig. I had no sooner written my brave words about sticking it out than I found myself walking ashore to make a phone call. To some friend, any friend, anywhere. I needed encouragement. Maybe I hoped someone out there had the power to talk me off the boats.

On my way to the dock's telephone, I ran a gauntlet of Baroid yard men who lounged by their dispatch office, a greenhouse on stilts. Five men, all silently picking their teeth after their supper, inspected my self-conscious breasts as I passed. Across the fence, occasional helicopters touched down from their last run of the day, driving minor storms of red dust up into the calm of a pink-streaked sunset. The Baroid forklifts and cranes had stopped cold for suppertime break under tall orange mud tank towers. Grasshoppers stung the hot silence between helicopters, and my footsteps were too loud, too sudden, on the dispatch office stairs.

I called one almost-forgotten number, then another. Long rings, no answers. No one out there for me. Ostracism is excommunication and can drive a person crazy, you know. That's what it's for. I would have cried out loud if it hadn't been for the Baroid dispatcher eating dinner at his desk across the room. I called my mother next, for proof that I existed. No answer there, either. The effort to pick up my jaw and screw it into a thank-you-for-the-use-of-your-phone smile was just too much. I hung up the phone and slipped out the door.

On my way back down the stairs, I found the Baroid boys absorbed in a real-life nature drama that stopped me too. A celluloid-orange hornet with iridescent blue-black-violet wings had stung a fat velvety spider to death. Now he hauled the stiffening spider, easily five times his own weight and ten times his size, from side to side, this way and that, attempting to shove it into a crack between two green clapboards.

The hornet lugged and tugged with his fore, tip-tap explored with his stern. All this action vertical, below the jagged crack. Not wide enough for an entry, it appeared. The hornet wouldn't give up trying.

Questions: Would he die trying? Isn't there some biological trigger that will stop this scene before it goes too far? Am I myself too driven to admit that my sea dream will never come true?

Half an hour must have passed while the hornet tried, tried, tried. Our human attention strayed. We talked of dumb animals, of dogs we'd owned who never learned their names, of monkeys unable to extract fistfuls of goodies from narrow-mouthed jars. I enjoyed this ordinary human small talk. It had been a week since anyone had talked with me at all.

The hornet labored on. Searched, lugged, wore himself down with his self-appointed task. Would some spark in me go out before I saw that what I wanted could not be?

The hornet dropped his catch once. But when a Baroid boy tried to help the hornet by retrieving it, the hornet attacked. Don't you dare interfere! I knew the feeling.

The hornet menaced us all, no doubt thinking we had designs on his supper. We retreated in a noisy herd. Okay, have it your own way, beast. And he did, making the long pull from ground level to eye level with that eight-legged grizzly bear a thoroughly dead weight.

We humans turned away to talk again. One of the boys produced a jay. A few of us smoked it. One that the rest called Old Man, because he was, asked to feel my biceps, a not uncommon reaction from oilfield men when they heard I was a deckhand. He felt mine. I felt his. Mine were bigger. Hm. When we turned back to check on the hornet he was collapsed atop a gas meter box, smothered under his burden. An irrational tear sprung to my eye.

But the hornet wasn't done for. He stirred, flattening himself to the meter box, extracting his limbs carefully. We cheered, all of us, when his last item of personal hardware was free.

And then we saw him think, actually think, and measure. He marched back over to the crack, alone this time, picked the

widest spot in it, paced it, worried at it, checked it out carefully. Then he quick-marched back to the spider. We more
discerning humans shook our heads, cried "Oh *no!*" Surely he
wasn't going to try it again. Not just when we thought he was
wising up . . .

Supper break was over by then. The humans paused, reluctant to walk away from the drama before its climax when the
hornet would either give up or die trying. But just then he
hoisted the spider up into the hole, climbed in behind it, and
was gone.

One of the boys hollered, "Geronimo!"

I went back to the boat then, and stayed out my hitch. There
had to be a way. I would find it.

Captain to captain, boat to boat. There were a couple of
good captains in there, one a man who played with porpoises.
Ordinarily stoic, Captain Rudy of the *Kingfisher* would seem
to be all business, navy style. I knew him for only four days,
but he was a welcome break from the tense, taunting captains
I'd grown to expect on each new boat.

Captain Rudy just didn't give a damn. Never did he mention
my gender let alone expect more or less from me because of
it. He seemed preoccupied. I know because it was my duty to
make up his bunk every morning that he kept a bottle there,
but even as fast as I saw it empty, he never took to drunken
ravings on the nature of manhood, womanhood. I learned
later that his wife had delivered an ultimatum. Either he came
home and worked a land job or she would divorce him. He'd
decided to go home. There were, apparently, more ways than
one to lose the boats.

Sunday was the crew's free day on Captain Rudy's boat,
and since we were tied up in shallow water to a workover rig
at an old pumping platform, there was the possibility of good
fishing.

I was practicing splicing line while Angel, our coonass cook,
pulled up a healthy batch of catfish. She snagged a mess of

even less desirable fish, too, and it must have been her tossing the trashfish off the boat that attracted the three porpoises.

Captain Rudy was musing alone at the stern, pulling on his sweet-smelling pipe, when he spotted them. In less time than it takes to say it, he put down his pipe, shucked down to his shorts, and arced over the side in a smooth dive. When he surfaced, he called back to Angel and me, "Watch what they do when I go dead in the water."

He struck out south, swimming hard, away from boat and platform into the silver-blue expanse of open Gulf. The porpoises, gobbling trashfish on the landward side of the boat, seemed not to notice.

"I'm drowning! I'm drowning!" we heard the captain call. Was he serious? Anticipating my confusion, he called out again, "Don't saaaaaave me!"

He struggled a moment longer, thrashing, blubbering. Then, sure enough, the porpoises appeared beside him.

Rudy, who had played this game before, went limp then, bobbing on the gentle water. One of the porpoises circled him twice, poking Rudy with a bottle nose. When Rudy didn't revive, the three sea creatures joined together to nudge the man back toward his boat, keeping his head above the occasional swells.

A tingle ran over me. I heard Angel draw in her breath. "Can you feature dat, *chère?* Can you feature dat? Dey savin' him!"

Nearing the boat, Rudy dared to stroke his rescuers' seamless backs and shining dorsal fins. They arced away. Rudy went limp again, sank a little, let out a stream of air bubbles. As quickly the porpoises returned to buoy him up.

Did they wonder why we didn't rescue the man ourselves?

They did, at last, present him to us at the open portside gate, pressing him to the boat's mildly rocking bumpers.

As many porpoises as I'd seen out there, I'd rarely glimpsed their faces. But here were one, two, three at once, upright in the water, all bland angelic eagerness, with smiles as wise as babies', awe-touched.

Angel gladly tossed them chunks of the single amberjack she'd caught and cleaned for our supper. I hauled Captain Rudy from the water. And then the porpoises went away.

Rudy was gone, too, the next day. A coarse, mean-mouthed new captain came to replace him. At first we sat in dock, peacefully, long enough that when we got the call to go off-shore, our blood was up for a run. Twigg, my deckhand partner, and I pulled the lines off the bitts, snugged them down on the stanchions, and climbed to the wheelhouse with the rest of the crew. The captain steered us out into the channel against a stiff south wind.

We watched a marsh hawk drift out of the high brush at bayouside and snatch a struggling meal from the squall-ruffled water.

Twigg was new to the coast, a Birmingham city boy, as easily amazed as once I was. "I guess he eats that fish alive, huh?"

"No diff'rent from eatin' pussy," pronounced the captain. "Bet it taste and smell about the same." He smirked, smug with carnal knowledge.

"You and yo' pussy mouff," our Angel snapped at the captain. "Look like wid dat ol' beard you got and not a toof in yo' head, you got a pussy *fo'* a mouff!"

Uneasy laughter spread, and Suggs, our fogbound alky engineer, oblivious as ever of the sense of our chatter, told us this story:

"Old Mexican, old old man, lived out at the edge of Bradenton where I grew up. Called him 'Hot Tamale' 'cause that's what he'd be eatin' every time. Used to go out there, bunch of us kids, see old Hot Tamale. When he died they took up the floor and found cat heads this high."

Our captain, whom Angel antagonized and called Puss-Mouff from that day on, radioed our office when we hit dock again. I heard him tell the personnel man it made him nervous having two women aboard at once. One was all any man could be expected to handle. This was Angel's regular boat. I moved on. From boat to boat, captain to captain.

I had developed by that time a strategy I hoped would guarantee my getting by the No Women on Boats rule: I would simply do what I was told, keep my too-smart lip buttoned to the last resort, and keep out of sight doing the dirty work no one else would touch. Sooner or later some sensible captain would notice my good work and keep me on.

Chipping down the invariably rusty rudder room, far below the daily social life of the deck, is the task I most often assigned myself. Chip, chip, chip. By the end of a steady morning's chipping I'd have a satisfying thirty pounds or so of rust to dump over the side.

Deckhands traditionally avoid rudder rooms because they're so small, cramped, never even tall enough to stand straight up in. Dim, too, until I hauled in my trouble light. But behind the steady thrum of the generators, I found peace in chip, chip, chipping. It was in my first rudder room that I discovered what is meant by "pumping iron."

It surprised me the first time, when I finished a solid hour of chipping and saw my right forearm plumped into a grotesque Popeye shape, giant veins a blue watershed in throbbing relief on my sunbrown hand. Aieee. Could this be my hand, my own arm? I thought I might have sprained it, or broken a blood vessel, and rushed upstairs to check with the cook. He said that an arm like that was what happened when you pump iron, nothing unnatural. "Keep that up, girl, and you'll have you an arm big as a boxcar."

Now I kind of liked it: a badge of hard labor. I went back to work, switched to chipping with my left hand until both arms were pumped up solid. Not quite as eerie looking that way. Chip, chip, chip, I worked on. Not that the work ever saved me for long from what I dreaded: the captains.

That one, on the *Cormorant*, grabbed my breasts, frankly, with both hands, when I passed him in the companionway. My next captain, aboard the *Ibis*, greeted me at the gangplank with "Well, I'll be goddamned. First they send me a nigger. Now they send me a woman! Tomorrow, you watch, they'll be sending me that nigger back again. Must be they tryin' to run me off."

Byrd Marine did send "that nigger" back again, not the next day but a week later. We were without a captain for two days.

Like all retired Caucasian civil rights marchers, I'm ill at ease with black people who aren't easily befriended. Such a one was Slow John the Shadow, our fill-in engineer on the *Ibis*. Black and glossy as bats' wings, clean as a new coin, deliberate as Bobby Fischer going for check and mate, silent as what they called him, The Shadow. He never spoke. Questions shot his way by the mate were answered with a sharp nod or a curious gesture that amounted to "I'm on my way": a cock of chin and finger thrust into the air. John was not deaf or mute, I knew. Why such silence?

At mealtimes, John filled his plate in the galley, without comment, and carried it to the back deck where he ate alone.

His working habits were as unobtrusive. Once he came aboard I noticed that the boat's minor mechanical problems vanished one by one. A light switch in the corridor that wouldn't switch suddenly switched again. A watery gauge disappeared, mysteriously replaced with a new dry one. A leaky soft spot in the main fuel hose got spliced out, somehow. It must have been the work of Slow John. Nobody else on that boat gave a damn.

In the heat of a delta summer, it's a rare engineer who will spend even half an hour in the blast-furnace heat and driving roar of the engine room. John, though, rarely emerged from that hell. Even our redneck mate acknowledged that John was "a good nigger."

After two days of John's silent treatment, I broke down, blubbering, babbling. "Talk to me, John. I mean, you must have seen a lot out here on the Gulf . . . and, to tell you the truth, I get nervous with you so quiet . . ."

Wincing at my unseemly display, John almost looked at me, but he didn't answer.

Next day, leaning over the rail near me while the boat moved down to the barge terminal, John spoke up about the bilge alarms, how he'd tried everything he knew but still couldn't get them to shut off when they were supposed to. Then he paused, and he paused.

I jumped in, blabbering again, to save the conversation. "How has it been for you out here, being black? The white men don't want your kind on the boats, or my kind, either. But you've stuck it out. How many years now? How do you do it?"

My blurry questions hung in the air for long moments. John paused, then yawned, stretching his salmon-colored palms over his head, then slumped over the rail again before he answered. "I like my engine room kep' up to where I can sleep on the deck down there. Be a week of work ahead of me before I can sleep right on that deck. Then that other engineer, that Adam, he coming back and just mess it up again. It ain't right."

I only caught his meaning because I lived in his milieu. He was saying this: that white men, rednecks, seem especially outraged at having to share bed and board with people of John's race. So Slow John eats alone on the deck, sleeps—imagine it—in the hellish din of the engine room, and keeps a silence that confers its own invisibility. I'd been looking for the way. There had to be a way. John had, he believed, found it.

John was fifty years old now, a ten-year veteran of the boats, and our brief talk depressed me, seriously. I'd been glorifying my own act all along, calling it Wonder Woman. I was only a good nigger, and not even a good enough nigger to keep a place on the boats.

23

Not all the men I worked with in the oil-fields threatened me, competed with me, wished to own me or at least humble me. Some softest part of the hardest, horniest guys sought me out, looking for a safe place to break down. I'll bet I saw more men cry that year offshore than I would have met with in a year of group therapy.

Richard, my partner deckhand on the next boat, the *Heron*, didn't cry, but he did show me his tenderness.

A profoundly ugly man, Richard was big enough to make two of me. From his narrow, rounded shoulders an outsize belly drooped, swelled, overflowed at the tightening of his belt. An inch below his hairline his brows grew together on a shelf that overshadowed close-set eyes, a nose upturned to confront life nostrils first, and a cherub-red mouth framing a knot of protruding teeth. Soft incurled ears sat at an Alfred E. Newman angle on his hubbard squash of a head. Alongside his brother, our captain, a man who could have modeled YSL for *The New Yorker*, Richard was a genetic joke.

Running out to the Gulf after a hurricane that had confined our *Heron* to dock for three days, Richard and I sat cross-legged on the steel shelf at the bow, relishing the speed, the rain-freshened rush of wind, the bloom of water away from the *Heron*'s hull.

We called out to a group of porpoises making crescent progress back to their fishing grounds from their upchannel storm shelter. Some few of them, one very small and perhaps newborn, peeled off from their fellows to leap in the rush of water at our bow. "I love them things. I just love 'em," Richard shouted to me over the wind. "They're half the reason I come out on this damn boat, just to see them."

I wondered aloud why porpoises keep company with human beings. "'Cause we try to be as good as them," Richard said.

Soon the porpoises went their own way. We humans hung tight to the forward mast to keep our balance as the *Heron* lifted and fell in the long swells of blue water. Forward like a merry-go-round, rocking like a cradle. We sat longer at the running bow that night than I had ever done before, sat through the kind of sunset where the orb is simply fuschia and the clouds crowd into the opposite horizon so as not to steal its glory.

When the darkness fell, with the wind almost too loud for talking, we pulled our jackets more tightly around ourselves and raised our eyes together to watch the stars pop out, paled by a rising three-quarter moon.

The sea, where our wake cut through it, fluoresced green-white. "They told me what causes that," Richard shouted, "is little tiny creatures. Tiny, tiny things you can't even see with your eye, but they glow in the dark. That's how you know they're there."

After a long silence, with both of us intent on the dreamlike glowing water, Richard spoke again. "Wonder what it feels like, bein' so tiny you're invisible. I always wanted to be invisible."

I could see him clearly in the dark, an overlarge and awkward hulk, wistful, gentle; Richard must have been a long time alone in his teens.

Late that same night Richard hammered on my stateroom

door. "Come see! Out on deck!" A sea gull leaned against our bulwarks, exhausted, probably lost from land and his flock in the recent violent storm. I imagined he'd been circling since, alone and disoriented, wearying, until finally he took refuge on this enemy, this boat.

We crept up on him and I snatched him into my jacket. A veteran of a hundred bird rescues, I felt the gull's feet. Cold. He didn't even struggle against me. In shock, probably, ready to die.

What I needed to restore him was an eyedropper of hard liquor and some warm milk. Richard went scouting, returned with a tot from the engine room bottle. I mixed the rescue potion, and Richard held the gull's beak open while I administered the dose. The gull was too far gone to fight back. He might make it; he might not. He needed warmth. On impulse, I handed him to Richard. "Here. You hold him."

Richard shrunk back, shy. "No, I couldn't."

"Look, I have to go to the bathroom," I lied. "Just hold him a minute, like you'd hold a baby. OK?"

By the time I returned Richard was the gull's natural mother. "He's warming up, Lucy, he's warming up. Here. Feel his feet."

So Richard held the gull, warming him with his own great belly while the brandy did its work. Then we wrapped the drunken bird in a towel, laid him in a bench box over a heating duct in Richard's room, and turned on the heat for the night.

We rustled up some leftovers for a snack and while we ate Richard showed me photos of his baby that he carried in his wallet. I admired the child, a ringer for Richard, ponderous, bald, and not at all pretty. But the grin on him, the grin of delight Richard had caught in one snapshot after another, vindicated the child's birth and breeding.

"I remember one time," Richard said, "I lifted Bobby up naked over my head, him giggling, couldn't have been more than three, four months old, and what did he do but shit in my face! I swear he did. Then he laughed and I laughed, and it just tickled us both. I wouldn't trade that one thing there for nothin' in the world."

Next morning I heard Richard knocking on my door again. Hell, it wasn't even 0500 yet. "Lucy! Lucy! The bird's awake and he's all better! Come see!"

When we pulled into our dock late that day the now lusty gull struck at our hands when we lifted him from the bench box. He fought Richard's motherly arms all the way up to the top deck. We stepped out into the glaring afternoon sun, and even the gull flinched from it.

"What do we do now, Lucy?"

"Throw him up in the air, high as you can. See those gulls over on the mud flats? Aim him that way."

"Just throw him?" Richard looked unhappy about that.

"Yes, that's all. Just throw him."

Reluctant but determined too, Richard deep-knee bent, clasping the querulous gull firmly with both hands, and rocketed him into the bright, cloudless sky. We squinted hard to keep sight of him as he lifted his wings in alarm, caught the wind, flapped once, then soared back in our direction.

"Look, Lucy! He's coming back! He likes us!"

But the soar turned to swerve, and we saw our bird dip the channel for a silver fish, then settle with his flock to digest it.

Richard had seen me writing, off and on odd hours crouched over my notebooks. Later that night he asked me if I would be telling about our adventure with the sea gull.

"For sure, Richard."

"Well, then, I've got another story you can put in your book or whatever it is you're writing."

"Fire away."

"See, once I called my wife from the Aces Bar. She hates that place, she really does. I only go there with my brother, but she hates to see me do it. Always a fight in there, those dudes offa the rigs and all. You know what she says to me on the phone when I call? She says, 'Well, long as you're there, just don't let the devil spit on your shoe.' Somebody ought to put that in a book. I never will forget it."

Everyone aboard the *Heron* was a dedicated marijuana smoker. On our trips in the Gulf we kept our eyes out for floating plastic-covered bales of it. One of the captain's buddies, a crew boat owner, had once fished half a dozen bales of marijuana—from an aborted Columbia-to-Louisiana smuggling mission, we guessed—out of the Gulf. One of our own crewmen had once found a bale on the beach near Galveston. Ours was not an impossible dream. Whenever one of us spotted a dark shape on the water that just might be a bale, we blew a special combination of hoots on the *Heron*'s air horns and called it Marijuana Drill.

We never did spot the real stuff, and eventually ran out of smoking material. The next time we stopped in Morgan City, Richard and his brother sent me over to the Vision Lounge, a notorious distribution center for dope, anybody's dope, with money enough to procure a bag. A lone woman would be less likely to arouse suspicion, or so went the theory.

It was early on in the evening when I got there, but already a gang of chromium hawgs and savage choppers clustered at the Vision's door. Inside, a nearly all-male population of deckhands, roustabouts, drifters, and bikers eyed one another ritually, not yet drunk enough to make something of it.

It was Friday, and Friday is the holy night of blood sacrifice in the Church of Blue Collar Virility and Recreational Violence. I hoped to be long gone before services got under way. Taking an uneasy seat at the bar, I ordered a straight double bourbon. Not my usual drink, but personal taste bowed to a greater need for backbone, social muscle. Besides, I felt a little dizzy with landsickness.

The Vision, it occurred to me, looked curiously like New York punk rock's CBGB's, and, come to think of it, just as much like the homo glory holes of the L.A. Strip. Curious, I thought. Matching congregations of black leather, silver zippers, heavy boots, cold eyes. The Vision could not have been built in imitation of CBGB's. The reverse must be true. Why so? Testosterone Culture is everywhere, I guessed.

I knocked back my drink, shuddering, and walked to the Vision's door and back, pretending to be waiting for a connection. On my second trip a drunken but familiar bearded face wove into my path.

"Don't I know you from someplace?" he asked.

For once this tired line was true. Here was the lazy deckhand I'd grown to hate on the *Harbor Pride*.

"Remember? It's me, Lucy. We worked together on The Goose's boat last year. We called you Bad Apple."

Apple's horny face hardened off into a sneer. He remembered now.

"Well, I got fired anyway," I volunteered.

Sneer dissolved into fraternal smile. "Me, too," he said.

We shook on it. I told him I was looking for dope. Had he seen Buddy, the blond dealer who could usually be found leaning on the Vision's front wall?

No, Buddy got stomped in the face, must have went back to Florida with his girl.

Then how about Trick, the cowboy who ran bags for that Puerto Rican dealer in Berwick?

Rick the Trick? He got knocked down so bad last week he had to crawl out of there. "I wouldn't say I knowed him if I's you."

Where was I going to get a bag, then? I placed the problem squarely on Bad Apple's shoulders. He took a run through the dark end of the bar and came back with news. A certain guy with bags of good Mex for forty dollars would meet me in half an hour.

Looking for protection more than company, I offered to buy Apple a drink, but he had what he called "blood business" across the road at the Pieces of Eight Barroom. When I sat back down at the bar I was glad to see another familiar face, RingWorm, a biker and a sometime oilfield engineer. He was wearing black leather over bare chest, studded leather wrist straps, dirty jeans with a chain from belt loop to wallet, and a neck brace, one of three plasters I'd last seen him with.

"Feeling any better, R.W.?"

"Well, now I got some trouble with my brain. They found out when I got wrenched in the accident, my brain got turned half around in my head. I get dizzy. It gives me a real bad nerve pain if I turn my neck. That's how come they put this damn brace on."

Still, his mood was good. He tugged me outside to meet his woman, Erny, who was pulling up to the door of the Vision in RingWorm's custom-painted death's-head van, The Ghostrider.

Erny operates a cherry picker for SubTech on Highway 90 in Morgan City, or did then. She told me she was thinking about going offshore as a roughneck on the rigs to make what she called some big buckers.

This could have been a significant moment for me, coming face to face with my working-class heroine: Tugboat Annie come to life, transposed to Morgan City. Trouble was, Erny —a twenty-year-old, slightly chubby blonde—had a black eye, another ugly bruise on her cheekbone, and a fresh tattoo on her starboard shoulder. It was RingWorm who'd bruised her, Erny told me, and the tattoo was his making-up present. An eternal rose for her shoulder, with "RingWorm" written in a romantic script on the ribbon running around it, and the day's date. The tattoo was seeping a little.

"Very pretty," I said.

Erny smiled. We talked a while.

I heard RingWorm, behind us, challenge a drunken youth who'd lurched too close to our circle. I'd barely noticed the kid, soliloquizing in our background, faltering ever closer, oblivious of our presence.

He was bruised, too. The text of his raving concerned itself with "how girls want passion even when they don't want sex," an interesting enough idea.

Then came the skirmish, quick and clean: a brief scratch of feet in gravel, a thud of startled human flesh into unyielding chassis, a sibilant intake of breath. RingWorm had the boy spread-eagled tight up against the thirteen coats of hand-rubbed lacquer on The Ghostrider.

"Don't go butt into conversations where you ain't wanted," R.W. menaced, close to the boy's ear. "That's one thing I can't stand, you hear me?"

The kid stuttered a promise to butt out and was allowed to stagger away. RingWorm returned to our company for a mouthful of warm tequila from Erny's hip flask. By that time I was so unnerved I forgot what I was doing there at all.

The drunken kid (hadn't he had enough?) called back a surly "Fuck you!"

RingWorm, satisfied that the kid was beyond Primary Defense Distance, kept his back to the boy and called over his shoulder, "Do, and you'll never go back to dogs."

R.W. leered at me then, as if to include me in his offer. "That I guarantee. That I guarantee."

Men. I would have said, and had said many times, that I loved men, that men were wonderful, men were exciting. I could no longer make such categorically enthusiastic statements. I'd walked unwary into their world and they'd spit me out again. They saw me as an alien; all right, I would study them from an alien's viewpoint.

The picture I put together was not pretty. Gruesome contests, barroom brawls, hell weeks, hazings, witch hunts, hierarchies, weapons, wars. Given that place and those men, given my furious innocence, my conclusion was inevitable: testosterone is the root of all evil. This simple and delectable construct saved me from facing a less orderly reality, for a while.

What struck the first wrong note in my two-legs-good/three-legs-bad system was what I came to call my Vision vision. Talking to Erny by the barroom door, I saw the shadow of her jutting hip and cocked-up shoulder reflected in the dancing-dark sidewall of RingWorm's van. I felt a blur behind my eyes, a swimmy recognition. Where had I seen this shadow before?

That vision haunted me, nudged me for a week. It was as if I'd left home on a trip and had a nagging, half-conscious feeling that I'd forgotten something. Something, something, but what?

Back on the *Heron*, more than a week later, at night, off-shore again, I was standing under a dim red port light when the Vision vision reappeared in a shadow just behind me. A ladder broke the shadow, rippled it, but couldn't disguise its pose of toughness. My own shadow, shadow of another tough cookie trespassing in the masculine idiom.

I'd been hearing the flat nontonations of redneckism in my own voice. I'd noticed my new juvenile hall posture of one clenched fist on jutting hip, the other in my jacket pocket, also clenched. Standing behind my face I could feel how it had flattened into a death-trap boredom I did not feel at all. It wasn't safe, I reasoned, for me to let my fear, anger, interest, excitement, show on my face, in my stance. Instead I'd taken to posing as a challenge: You Wouldn't Dare: playing at a game of toughness that I could never win.

That question again: Who did win?

Plenty of the men were giving me a hard time in the oil-fields, and the worst was yet to come. But I was beginning to think I would someday have to forgive them for it, and the Vision vision is what started it all.

24

The Heron *was my haven* for a while. The boat herself was sound and nicely made, the crew was friendly, and Richard's brother was a good enough captain. Good enough that another boat company hired him away, and Richard too. When they left, my new captain, Aubrey (pron. o-BRAY) Mouton, came aboard. Aubrey was my twentieth captain; I was the first female deckhand he'd even heard of. He was deeply outraged by my presence on his boat.

He could simply have set me ashore and called for a replacement of the appropriate gender. But that wouldn't have satisfied his apparent need to debate me, refute me, prove to me once and for all that I was dislocated, deluded. He had an axe to grind.

"Women are like horses," he gave it to me straight on our first meeting. "They can be just so kind, lovin' you up one day, next day they turn up their noses when you rattle their feed bucket."

He ought to know; he was the captain. He was the captain; I was required to agree.

Maritime law invests each captain, young or old, sea-wise, incompetent, or psychotic, with perfect omnipotence. In a word, the captain's word is law. A captain need never offer even the shakiest rationale: "Because I say so." This much is universally understood.

"The captain's word is law" has been the key line of a womanjoke in our family since I was thirteen. It refers to a brass plate conspicuously displayed in the home of our family friends, the Sinclairs. Bill Sinclair, in theory head of household, was outnumbered and outmaneuvered on all sides by his daughters and wife. The Sinclair family culture was no matriarchy, more like an anarchy by virtue of the female line's easygoing carelessness. Bill was the sole hysteric. "Nobody listens to me! Do you hear me!? Nobody hears a damn thing I say!"

When Bill bought a pretty little Chris-Craft, he planted a skipper's cap on his watermelon baldness and nailed a brass sign affirming his supreme authority to the wall over the family television: THE CAPTAIN'S WORD IS LAW.

Ruth Sinclair told my mother that, sure, the plaque was foolish, but she wouldn't dream of removing it. "It's kind of a man's touch. It's *nice* to have a man around the house."

Out on the blue water, a captain's word cannot be so blithely dismissed. A captain may outlaw black pepper from the galley, or insist on the appearance of beef (O beef, manly beef) at every meal. He may send his minions out into the teeth of an ice storm to rinse down the boat, or frown on their least ambition. And God save the crew in a political year. If the captain says that the incumbent has sold out to niggers and commonists, he's got maritime law zipping the lip of conflicting opinion.

As one oilfield mariner put it, "Five million cap'ns in this world. Can't all be God." Try telling that to a captain.

If it were up to him, Captain Aubrey told me, all women would be home where they belong instead of going out trying to compete with the men. But the world was getting out of control. It was coming to where women and niggers thought they could take over.

He said he intended to be fair to me ("I won't cut you no slack, neither") and allow me to prove myself if I could.

This "proving oneself," I learned, is an ongoing process. There is no happy ending to it, no arriving at the point where oneself is finally proved. An example:

"Betcha can't do *this*. Couldn't if you tried."

I did it. I'd been doing it all along.

"Yeah, but I betcha couldn't do it in side seas."

When they came along, I did.

"Yeah, but you did it with the light one. Wait'll you try it with that heavy one we keep in the paint locker."

I did it with the heavy one, handily.

"Yeah, but you couldn't do it twice in a row, with the mud on it."

I did it three times running, with mud on it, in a vicious wind.

"Yeah, but you was breathin' hard. I seen you."

Eventually I did it effortlessly. Came the inevitable unanswerable: "Yeah, but you can't do it and remain a lady, I'll tell you that."

It wouldn't have won me the point if I'd done it wearing white gloves and a delicate veil over my picture hat. Because the captain had an unlimited supply of unanswerables:

"Yeah, but you couldn't never of done it slick as I did when I was deckin'."

Or: "You want to get big ugly muscles on you, want to ack like a man, thass your bidness. Want to lose any chance you ever had to git yourseff a husbin, go right on ahead. No skin offa my butt."

Or, finally: "Don't let me see you do that agin."

Proving oneself, then, is learning to live on when victories are snatched away, to proceed "manfully," as though the next victory will be real and final. If you do manage, miraculously, to accomplish some act of climactic proof, you are "just trying to prove yourself."

But Captain Aubrey was no simple John Wayne. There were holes in his bucket and he knew it.

We sat some long watches alone together (in fact he seemed to prefer my company to any other aboard), and it was in the confidential darkness of the wheelhouse that he listened as best he could while I spun out my alien philosophies, my dreams and ambitions so foreign to his image of womanhood.

"Thinkin'," he told me once, "now there's somethin' you're real good at. Thass what it is about you: you think like a man."

Of himself, Aubrey said, "I'm not as bad as I used to be." His wife had had enough of his acting the sea captain of their home and had hauled him to a marriage counsellor. In the darkness of a wheel watch I heard the snuffle of his tears. He'd come close, and might again, to losing his home, his woman, his sons.

Reviving from that memory of despair, with an ironic edge in his voice, Aubrey baited me: "How come you want to push into a man's job, anyway? Kin't you be happy, have a nice little family, go to church on Sunday, give the old man a roll in the hay onct in a while, keep him happy, huh?"

He knew, he knew.

"You really like this old *Heron*, don't you?" Aubrey had seen me snugging up her lines lovingly, patting her battered old hull, leaning a cheek on her stanchions to watch the stars emerge from twilight. The boat was a lover we shared.

When evil weather came we rode it out together up in the wildly pitching wheelhouse while the rest of the crew mewled and retreated to quarters. I'd catch his eye catching my eye when a particularly violent roller lifted the *Heron* on her stem and spun the wheel in Aubrey's hands. In those brief and precious moments we acknowledged a comradeship. He liked my spirit. I liked his.

On another evening Aubrey and I stood on the back deck watching the chromatic progress of a sunset. I told him how often during my nights ashore I'd dreamed of boats leaving without me. How whenever I got back to sea again, those dreams dropped away. Why the change, I wondered.

Aubrey grasped the meaning immediately: "Oh, sure. You were missing the boats."

Having puzzled over and just plain failed to catch that obvious dream pun, I leaped on the jolt of Aubrey's understanding to a wide-eyed realization. I wasn't all this time just saying I loved the boats; I did love them, wholeheartedly, desperately. If this running away to sea were merely an *idée fixe*, it was certainly fixed deep.

I felt a whoosh of release, of pure joy. I danced a silly circle around Aubrey, who looked on indulgently. He said he knew just what I meant and how I felt. "It does," he said. "It gets in your blood."

Now tell me something; maybe you can figure it: Aubrey knew now that I was a sailor in my heart. He saw every day the evidence that this was my true calling. I respected his captaincy, I loved his boat, I did them both proud. Why did he hound me off the *Heron?*

Public occasions found him dead earnest at his captain's role, found me—and I should have known better, too—fully engaged in running his treadmill of testing, testing, testing. Every day Aubrey relied more heavily on the efficacy of public and fatherly chats. He shook his head sadly over the loss of my womanhood, railed at women everywhere, foresaw the day when I would stab him in the back or let him drown. Disputing his prejudices only escalated the psywar. What might begin as patronizing counsel could end in the ultimate macho pinhead challenge: "By God, what was good enough for my mother is good enough for you!"

I believe I know, now, what men ask when they say that: "Do you refuse to suckle me?"

I wasn't half so smart then. I rode the waves of Aubrey's psychic storms alone and confounded. By day he challenged, backed and filled, claimed to cherish, played lord of the manor, played little boy lost, raged like a wounded Zeus. By night my own dreams hammered at me: father is dying, you must help help help help . . . father is angry, you must run run run!

Every workaday job was another ultimate test. Could I lasso the dock's bitt from our moving deck the first time, every time? Could I stack the three-hundred-foot length of muddy anchor chain without a rest stop? Could I build muscle tissue fast enough to keep up with the parade of tough young deckhands brought on board to show me up?

My center of gravity moved from belly to throat; my throat started aching. I was first dizzy, then fully ill. I passed out while driving myself to clean a mud tank in record time. A helicopter came for me, lifted me to an onshore hospital. I did not then see the psychosomatic connection.

When I returned to the *Heron* for my second hitch, Captain Aubrey met me with an ugly skepticism. "Soon as the going gets tough, you'll git 'sick' again."

I just growled and buckled down to prove myself some more, chipping the *Heron*'s rust spots down to her gleaming steel, putting a shipyard finish on her house, splicing out the frayed sections of her lines. She became my obsession, my pride. But still I couldn't win.

Captain Aubrey's shadow fell over me one morning as I scrubbed the indoor stairs. "You'd make some man a good slave," he smirked. "Why don't you just come on and crawl into my bunk."

I gave him five seconds of silence, a shot of volcano eyes. He didn't flinch, wouldn't back off. He'd been drinking.

"Y'know, I expeck women on my vessels to obliiiige me," he said, stringing out "oblige" to a near-yodel.

So this was it. The battle line was drawn, the challenge made that must be answered. I didn't stop to wonder what's a good nigger to do. My anger boiled over.

"The kind of 'woman's work' you're talking about comes a little higher than forty-five dollars a day, Cap," I spat at him. "Why don't you grab one of your skin books and go obliiiiiige yourself?"

Aubrey's answer was emphatic, phallic, a territorial display of the first water. He stepped over me into the bathroom at my elbow and took a thunderous piss. With the door open.

For the remainder of the day Aubrey lurked in his stateroom, as concentratedly quiet as the Mad Bomber at his workbench. Something was up, I knew.

When I came in from the deck for dinner, I saw what. Aubrey had put in the day scissoring juicy clips from his

Playboys, Hustlers, and *Rogues.* The galley was plastered with a montage of glistening pink vulva from wall to wall. I sat down to dinner, pretending not to have noticed the change in decor.

Over pot-roasted beef, Aubrey had this to say: "I seen you climb that mast yesterday, change that bulb all the way up. And I says to myself I might's well aks as go on wondering. Tell me, did you have a sex change? And if you did, are you going from man to woman or woman to man?"

My eyes were tired of stinging. I left the table.

The captain had bought a couple of cases of Budweiser ("the King for the King," he said) and it was under the influence of the first of them that he called me back to the galley for "a nice little chat."

"I'm sorry," he said, blurrily, "but it just happens to be true. Women can't be trusted."

Like horses, like cars, like children, Aubrey claimed, you had to put more into them than they would ever pay you back. Ungrateful, the cheap jezebels had neither the loyalty nor the sense to honor the side their bread was buttered on. Did they think it was easy, being the boss? Didn't they know a man couldn't go up against the whole world, then come home to whining and sassback? And answer him this: How's a man going to be masculine when women refuse to be feminine?

But one of these days they'd see it clear as anything. No matter how hard they tried, no matter how much of a brain they had, they'd learn (here fist struck table, crash reverberated in steel-walled galley) it's Brute Force that counts.

Halfway through the second case of beer, Aubrey raised his weaving head off the tabletop to say, "Oh, you know I like you. I allys did like you. You a funny bird but in a way you 'mind me of my mother. My mother, she's a good woman. She don't take shit offa nobody."

Still later he trailed me to my stateroom. He'd written a song in his head and wanted to sing it for me.

"Here it is: 'My child is crying . . . my girl is dying . . . They made me a man so I'd see where I am . . .' That's all I got so far, and I don't know what it means exactly, but it's a good song."

The next day hostilities took up where they'd left off. Aubrey barged around red-eyed and badly hung over, with his beardy jaw set hard, marking the boat with uro-territorial trails, breaking his black silences with ugly challenges:

"Women only good for one thing. Like I say to my wife when I get home, 'Spread your legs and earn your livin'.' "

"Sure," he leered, leaning over me, "I've had many nice little affairs with women cooks on my boats. Any of 'em give me some backtalk, I tell 'em pack up and git that fast."

The next morning, he fired me. "Terminated" me, as the boat companies call it. The charge against me? "Undermining the captain's authority."

25

Fired again. It was all my own fault, I knew. Sure, some of these captains were ignorant and mean and even crazy, but I should have been able to live in peace with them. The other hands did.

More than ten months had passed. I'd had twenty captains. Twelve of them had rejected me. I couldn't pretend that I was doing right and twelve captains were not. Clearly, I was mishandling my situation. Failing. And not because I was a woman but because I was something worse: myself.

Maybe if I stuffed my pride, let the men do my job?

Maybe if I played this game for what it was, a business, if I kept aloof, the men might respect me for it?

Maybe if I slacked off on the work?

Maybe if I went to bed with a captain once in a while? Captain Aubrey was attractive enough; I even felt a kind of love for him on those stormy nights in the wheelhouse. Maybe I didn't have to be a hardnose about refusing sex?

Maybe if I pretended to be a dyke?

There had to be a way. I just hadn't found it yet.

Once in a while I considered that the problem did not lie with me, that it might well be insoluble. I brushed that thought away. Every problem has a solution. I was strong; I was smart. I would find it.

Facing Guste again, delivering up one more link in my chain of failures, was hard. I saw a shadow pass over his face and thought it meant: not *her* again. But no, he was only pissed, at Byrd Marine, at my captains, at the gossip about me on the docks. His own Watercraft Company had withdrawn the offer of the cook's job; I was too well known, too controversial. That crazy Yankee morphadite, a shitload of trouble with a loud lip on it. "Ah don' know, *ma fille,* what we gon' do wid you." Guste shook his head sadly, rocking me under his arm.

He gave me the name of his contact at Petrolco, a major oil company. Maybe Petrolco's man was too far removed from the docks to have heard of me yet. I could try for a job there.

The nervously hearty Petrolco man conducted me to his office, "although I'm afraid I can only give you disappointing news." His company planned to have an affirmative action program for women on boats by 1981, two years away. A pilot program to employ women on their oil rigs was just beginning, but the quota was filled.

"I'm glad to hear you have a program, anyway," I said. "How many women are you putting to work on the rigs?"

"Six, actually. Three of them are Native American women from a reservation in Texas." He put a sanctimonious inflection on the word "reservation."

I was shocked. "Six? Six? Is that one for every hundred billion dollars you made in profit last year?"

Petrolco Man flinched. "But we'll have your application right at the top of the pile if one of our experimental ladies can't hack it out there."

I'd had enough of being an experimental lady. I said no thanks and split.

I stopped next at the live-oak-shaded offices of a solid old firm with a fleet of tugs and supply vessels. Sadly enough, the receptionist wagged her head, they'd just that moment run out of application forms. But if I wanted to leave my name?

Seething, I retreated to the Sunshine Rib Shack with a Morgan City help wanted section. The ads, I saw, were separated into Help Wanted Male, Help Wanted Female. There were two entries under Help Wanted General: one for a florist, another for a poodle groomer, jobs presumably reserved for the indeterminate of gender.

Half the blockbuster display ads carried the *Equal Opportunity Employer m/f* line, but they were clustered on the other side of the Help Wanted Male wall. I thought I dimly remembered that such walls were no longer legal. I called the newspaper office.

"Not as far as I know," the *Shrimp and Petroleum News* publisher answered, hanging up. My imagination bubbled over, picturing his rheumy blue eyes and misshapen wedge of nose technicolored by bourbon-blasted capillaries, his Hitchcock lips outthrust in shrimp and petroleum satiation. Yes, that would be my man, the enemy, and he was winning. If I got to the boats and then failed, it would be my own failure. Being prevented from getting to the boats was something else again. Oppression, maybe.

Two stops later (at one of which the employment secretary confided that I was *welcome* to an application form, but her employer would never call a woman for the job. How did she, a woman, feel about that? "Personally, I don't know *what* kind of a woman would want to go Out There."), I was ready to howl at somebody, anybody.

I found the office of the local congressman nearby. (He has since been elected governor of Louisiana, interestingly enough.) His female aide composed her long white hands on her mahogany desk and inclined her well-groomed head sympathetically. I explained my mission and she stopped me there with an imperially upraised hand.

"Tell me something before we begin, Miss Gwin. Do you believe it can work?"

"Do I believe what can work?"

"Women working offshore, on the boats. I understand that the most vigorous objections to it come from the captains' wives."

I bolted up from a terrifying dream that night, having been awakened by my own voice. Deeply, coarsely, monotonously, it was saying, "You gotta fight 'em. You gotta fight 'em."

Next morning, on the off chance that I might have legal rights in this head-on clash with Byrd Marine, I hitchhiked to the local college library and studied what little material was available on civil rights and sex discrimination law. What I read shocked me. I had rights, actual legal rights. Sexual harassment wasn't just some trendy feminist term but the legal name for an illegal act. During my months on the boats, in ways I hadn't fully understood, Byrd Marine and its employees, for whose acts Byrd was responsible, had prevented me from earning my living. Even the everyday galley table gang-up-on-the-woman game was illegal according to the guidelines of Title VII of the Civil Rights Act. Wow. My country was behind me, all the way. I didn't have to go on fighting my battles alone. I could get a lawyer, go to court.

I spent seventy-two dollars that day on toll calls to lawyers, just about every lawyer in the Yellow Pages of three parishes (counties). Each one refused the case. Each one had a boat company client, even if it was only an oyster boat client. Representing a plaintiff in a civil rights suit against a boat company, any boat company, would be a conflict of interest, these lawyers believed. One of them referred me to a crony in New Orleans who was rumored to have brought suit against boat companies from time to time. Just damage suits, the New Orleans lawyer said. "I wouldn't touch civil rights with a ten-foot pole."

"Does that mean," I asked, "that maybe no lawyer will represent me in a civil rights suit against a boat company?"

"Don't ask me what it means," he said.

It appeared that I'd be preparing and pleading my own case. Hm.

Well, I consoled myself, it wasn't as if I wanted one million billion dollars in damages, or even revenge. I wanted my job back. Even a jailhouse lawyer ought to be able to manage that.

I had some slight advantage over the real jailhouse lawyers. I'd once worked as a legal secretary; I knew the language and the forms. I could spell. And since I wasn't locked into a cell, I could hitchhike to the nearest law library. I could even track down my witnesses. I hoped they'd be willing to side with me, but I doubted it.

I was lucky. Two old shipmates who'd been fired from Byrd Marine because they complained about the way Captain Aubrey had dealt with me were only too happy to sign affidavits to that effect. They even did some private-eyeing for me, carrying a hidden tape recorder into Byrd Marine's offices. They taped one personnel officer boasting about how neatly they'd gotten rid of me. "She wants her women's libber stuff, let her go back north and scream her head off. Won't do her no good around here."

Two other men I'd crewed with signed statements for me corroborating my own affidavit. I even located a Byrd ex-employee who told me I'd been underpaid by ten dollars a day during my months with the company. So far, so good.

In a matter of days I had enough of a case prepared to start firing off letters. I wrote to every organization and arm of government that might conceivably be of help: the American Civil Liberties Union, the Civil Rights Department of the Maritime Bureau, the Equal Employment Opportunity Commission. I even wrote to the top executive at Byrd Marine. A few days later an E.E.O.C. lawyer called me to say that my formal complaint was now on file. Soon I'd have some action.

Meanwhile, I was desperate for a paycheck. I went job hunting.

I went so far as to apply for the job of area supervisor with a Louisiana fried chicken chain. But I guess I'd lost the executive look because the boss chicken never called me back.

Every day I applied to one or more of the hundred and some boat companies in the oilfields, with predictable but still infuriating results.

"Kin you cook? Why don't you wanta cook?"

Or worse. "You ain't that morphadite I heard about, are you?"

I don't know if I felt relieved or not when a pipeline contractor took me on as a laborer. Four twenty-five an hour for odd jobs around the oilfields. I told myself it was temporary anyway. I was winning my way back to the boats.

On my first laboring day my foreman issued me an unwieldy implement called a ditch bank blade and set me to the job of cutting back brush from the fence around the Hemmoil dock at Cocodrie, Louisiana. Cocodrie for crocodile for alligator, which was once that end of the world's claim to fame until, according to local shrimpers' supply store owner Elmo Cavalier, "Dem Mississippians come and clean dem out."

Swinging that hard, dangerous blade under the September sun with the fire ants running up my left leg and my right leg thigh deep in the evil oily sludge that once had been a bayou, I had time to wonder what in the world had brought me to such a pass. For the very first time since I'd quit the advertising business, I wished I could sneak back to my expense account, my two-hour lunches, my twenty-fifth-floor view of civilization.

Next day I washed cars at the contractor's office. When I did a Wonder Woman job of it, the company's El Presidente decided he could trust me to scrub a grease spot off the carpet of his private plane. Then he brought his daughter's Continental to me, "for the woman's touch," to see if I could remove the poodle hairs from its cushy upholstery. I remember my stinging resentment, harder to conquer than my hatred for that ditch bank blade.

Have you ever sworn you'd rather dig ditches than . . . ? For the next two weeks, I had my chance.

If any job in the world deserves the label "man's job," let me nominate ditch digging. To bite with a dull shovel a patch of ground sun-baked to the consistency of earthenware, to fetch up heavy red clay hour after hour after interminable hour, to mark your progress in aching inches while the rest of the world whizzes by gawking at you from the comfort of their air-conditioned cars, that's ditch digging.

Let me assure you that no ditchdigger will ever write a novel in her spare time. Complex thoughts and trains of association do not occur. There is no time but work time, no will but to endure. At the end of a ditchdigger's workday, the only release for a ditchdigger's rage is a six-pack of beer and a shouting match with Walter Cronkite. Believe me.

I worked out my own *zazen* of ditchdigging, to sidestep the drag of time, the surges of impotent anger, the tides of encouragement, discouragement, fantasy, rage. To Just Dig.

I surprised myself on the second day when I climbed out of my nobly square-sided trench to see that my partner, digging from the opposite boundary line, was yards behind me. His trench had dwindled from four feet in width to three. Its walls were shabby, round. I saw, at long last, that there is no ultimate man's job.

But if they want that one, let them have it.

After ditch digging came a pleasant week of driving the contractor's light trucks. Here was my first real opportunity to tour the backwoods and bayous of the state where I'd lived for almost a year. Nice.

I took an unauthorized detour one day to New Orleans, to meet with an E.E.O.C. counsellor who'd be helping with my case.

Mr. Ortiz was a sad old doggy-looking fellow, a native of Puerto Rico, and after he'd reviewed my file he told me his own tale of disappointment. He'd come out of the barrio and put himself through law school hoping for a someday judgeship. "But you see where I am today, fifty-five years old and I will never be a judge." Then he refocused his faraway eyes, leveled them at me, and said, "Miss Gwin, we must all face Reality someday."

"That's never been my strong suit, Mr. Ortiz," I told him, volcano eyes rumbling.

Taking my suit through the federal courts could drag on for two years, four, maybe longer, Ortiz said. But if I settled out of court with Byrd Marine, abandoned the idea of winning back my job, the thing could be over in six months or less.

I shook my head. I wouldn't stop fighting until I had a boat under me again. That was final.

Then it was Sunday and my telephone rang. Byrd Marine's number one man was calling.

In his answer to my first letter he claimed he couldn't "defend [sic] our employees from repeatedly making a pass at one of the few women in an all-male environment. I feel that making a verbal pass, without conditions thrown in, should be interpreted as a compliment." As if Captain Aubrey had presented me with a bouquet of violets and invited me to dance.

I'd fired off a second, spikier letter telling the Byrd executive that I'd sunk my teeth firmly in his corporate pantleg. He wouldn't be shaking me off. To clinch that threat, I enclosed a copy of my complaint to the E.E.O.C.

Now the solicitous phone call. Byrd Marine was willing to hear me out. "I enjoyed your letter," he said. "You've got quite a sense of humor." I refrained from growling and we set up a meeting for that afternoon.

I will call him Captain Slicko, in honor of his Lacoste golf shirt and his kidney-shaped desk. He was likable enough, a Yankee anyway. He confessed, disarmingly, that his wife had read his letter to me and called him a turkey. Two of his young nieces were working as cooks on the company boats, so he knew what I was up against. He'd come up through the ranks himself. But he wondered how I could hold him responsible for what happened Out There when obviously he was occupied with pressing corporate matters.

I examined my fingernails while he preambled. Weak, I saw; he's weak.

When he finished, I launched into a semiprepared, heavily ironic chronicle of my experiences with his captains, his personnel men. Captain Slicko's face crumpled into resentment. I roared ahead, building momentum. My anger had, at last, found a worthy target.

"Your turn," I said, much later, when I ran out of breath.

The men, he countered, no doubt harassed me because of my abrasive personality. From where he sat, it looked as if I'd been *expecting* trouble with Captain Aubrey. No wonder I'd found it; I'd brought it on myself.

My answering outrage ran the gamut from common misconceptions concerning role and gender (with anthropological footnotes) to citations of maritime and civil rights law.

"You can't legislate morality," Captain Slicko parried.

I counterthrust with the Bill of Rights, America's original morality legislation.

Captain Slicko backed off then, conceding that his company had expanded too far too fast for him to have kept up with current events like the Bill of Rights.

As I've said, he was weak. A good target but not exactly a worthy opponent.

He'd been forced, he pleaded in his defense, to hire whatever men were available. Could he be blamed if they were madmen, bad men, thieves, morons, murderers?

Besides, he could see that I had what he called "a bad attitude." Obviously I believed myself superior to his boat crews (the above-mentioned murderers, morons, and thieves).

Gentleness, femininity, tact, and good home cooking would take me further on the boats than all the so-called rights in the world, Captain Slicko said. "It's men on top here and, unfortunately or not, it probably always will be."

Further, he wanted to see me prove myself on the boats, both physically and socially, before I came crying to him. "I like a real survivor," he declared, enjoying the ring of that word.

"Survivor? Isn't that what they call those goofy old men who hide out in the woods eating squirrel heads and waiting for doomsday? Don't you dare ask me to survive with you driving your kidney-shaped desk right over me."

The two of us sat back in our chairs then, breathing hard, red around the eyes from the pressure of applying semicivilized behavior to a crude problem of brute force. I realized that outtalking this white-collar sailor was not the point. The point was to win back my job and some measure of company protection.

I suggested that we switch off generalizing and get down to cases. What would Captain Slicko suggest I do when a captain, whose very word is law, makes a grab for my boobs?

Funny I should ask that. Because just that week his niece, my former friend Karen, had asked the same question. She'd been bending over the oven on her boat when a captain pressed his Levi-sheathed hard-on into the crack of her ass. "I told her I would have turned around and decked the guy," Slicko swore. "Of course, *she* should never do such a thing."

"Then what should she do? What did she do?"

"She gave him a dirty, dirty look."

So much for my volcano eye armory.

"So there you sit, behind your desk," I sighed, "telling me to go out into the vicious world of the offshore oilfields with nothing but dirty, dirty looks to protect me. It's just not enough. You have to do something too."

Captain Slicko coughed and admitted he needed to explore these issues more fully. He'd even look into my personnel file "to see what can be done." I asked if I might look into it with him.

And there it was, clipped to the front of my file: the real reason why Captain Slicko had decided to see me. It was a letter from the E.E.O.C. notifying him of my suit and ordering him to answer a lengthy form-in-triplicate questionnaire on Byrd Marine's hiring practices.

"Oh, this doesn't belong in your file," Captain Slicko said, snatching the E.E.O.C. documents away. Revealing, beneath them, something even more interesting: my report cards, a stack of evaluation slips, one from just about every captain I'd worked for, all of them praising my work as a deckhand. Straight A's, very nearly. I got goosebumps just thumbing through them. I hadn't just imagined that I was a good deckhand; here was proof.

The comments portions of the report cards were the most revealing. "Send me two like this," one captain said, "and I could paint this whole boat in a week." "First class on the lines," another said. "Too bad she's a woman," one lamented, and that said it all. I hadn't been drawing trouble to me. My failure on the boats was maybe not my failure at all. Whew. Wow.

I had a lump in my throat and tears in my eyes. Here, buried away in the office, was the acknowledgment I'd been breaking my heart to get. Captain Slicko, reading over my shoulder, purred. "There may have been some miscommunication here about 'undermining the captain's authority.' I'm all in favor of giving you a clean slate."

"You mean I can have my job back?"

"The first available deckhand slot."

I asked him what he planned to do about the two months' pay I'd lost. On this point he backed and filled. He'd have to check with the parent company. They didn't like setting precedents. "I'd want to be able to tell them you're wiping the slate clean, too."

"You mean I should drop the suit? Call off the Feds?" There were, I knew, other women pounding at Byrd's door for jobs and being turned away. I'd heard the cynical personnel man tell a captain that since he was supposed to put a black person and a woman on every boat, he was looking for "big ol' black mammy cooks." Drop the suit? I said I'd think it over.

Captain Slicko and I parted with a clammy handshake. He

had a piece of advice for me, too. "I had troubles myself, you know, when I was working my way up on the boats. Hell, I was just as much an oddball as you. I'd like to pass on to you my own trick for surviving on the boats, and I can give it to you in two words: Play dumb."

I noticed then what I'd been too preoccupied to see before: a handsomely hand-lettered sign on Captain Slicko's office wall. It read: "If you can't dazzle 'em with your brilliance, baffle 'em with your bullshit."

All the same, I'd won my reinstatement. I went home to pack.

Part Five
Going Overboard

26

"*So there's my* Wittle Wucy! How are you, darlin'?"

Captain Mick bit my neck affectionately and nuzzled my bare toes with his. Surprised to see him standing outside Captain Slicko's office, and even more surprised by the rush of love I felt for him, I hugged him hard and then stepped back to admire his "Think Crude" T-shirt, an oilfield souvenir I coveted.

Mick had made a special wardrobe effort for this visit to company headquarters: seersucker bermuda shorts, a flowered permapress cowboy jacket, and, incongruous in his normally toothless face, a perfect pair of plastic choppers. Shoeless, Mick's feet were nevertheless so clean that his thirty-carat bunions shone like spotlights.

"We sure do miss our Wucy, surely do. And now they tell me you're comin' back to your old buddies on the *Condor.*"

The hair on my scalp lifted, of itself. My face froze; I could feel it cracking. The *Condor.* Not the *Condor.*

I'd had a plan, to prove that with only a little help from the office I could make and keep a place for myself on a Byrd Marine boat. In just one more month of sea time I'd qualify for the Coast Guard designation of able bodied seaman. I'd be a certified experienced sailor. And since every certified boat was required to carry at least one able bodied seaman, Byrd would actually need me. My job would be safe. That had been the plan.

I could have marched into Slicko's office and refused to board the *Condor*. But I'd already insisted on having the very next assignment. The *Condor* must have been it. Hoist on my own petard, I think they call it. Must be a nautical term.

I held my breath and let it out again. I'd go quietly.

"They tell me we gotta be nice to you on account of some federal court papers you got," Mick said, with a flicker of evil in his eye. "I told the boss man we're always nice to our Miss Wucy. We'll get along just as sweet as can be, won't we darlin'?"

A helicopter took us out to the *Condor* that same day. My heart sank when I saw her. I'd painted her inboard a few months ago; no one had lifted a brush to her since. The paint on her outboard hull was patchy, peeling. It was late October now, and somebody had to get her ready for winter. Paint, paint, paint, here I come.

Gitzy, SpaceMan, and Danjones were crouched around the galley table stropping knives when we got there. Gitzy's bloodshot eyes widened when he saw me. "Lookit, it's old Wucy!"

I asked where was the cook. I'd hoped for another woman aboard.

"They won't send us no more cooks since we tied that last one to the P-tank ladder," Mick said.

"Spread-eagled her, too, didn't we?" Gitzy cackled. "Spread-eagled her and spread her all over with Wesson Oil."

"She had to go cryin' to the office," Danjones added. "Tryin' to get us in trouble. She was the one got fired, though."

I couldn't tell if they were lying; I hadn't heard that story around the office, and SpaceMan wouldn't meet my eyes.

I stood back while the boys traded welcoming wrestles with Mick. Three oilstones gleamed on the galley table, and a strop hung on the doorknob to my room. Every kitchen knife aboard had been honed all the way to the hilt on both edges, converted to what Gitzy was calling "stabbers." Jesus. Fourteen days of this ahead of me.

"Look what I bringed you, sweetheart," Mick said to Gitzy, producing a big old blue-black gun from his seabag. A cannon, more like.

"A forty-seven Magnum." Danjones whistled, delighted. "Blow a hole clean through our starboard main if you brace yourself right."

Gitzy turned the treasured evil thing in his hands. "Needs a little cleanin', Mick. Where'd you get it?"

"Took it off that nigger owns the barroom on Front Street." Mick grinned waggishly, toothless again. "Got you some amminition, too." And there they were, bullets almost as big as deer slugs. Jesus.

"Bring any booze?" SpaceMan wanted to know.

"I didn't forget you, old buddy."

I went shivering past the razor strop into my room while the boys cracked the first bottle of the party. Fourteen days.

Painting the long walls of a boat's exterior is the next best thing to a steeplejack job. Theoretically two deckhands are present at all times, one hanging over the side on what is called a Jacob's ladder, a handmade manila swing with its low ends dangling two stories down to the water, its high ends tied around the mast, or a bitt, or whatever else is handy and solid. One deckhand paints while the other stands by with paint refills, tools, and a pair of life preservers, just in case. After a while they switch jobs.

But SpaceMan ignored the whole operation and hung out in the galley with the others, polishing the gun, honing the knives, taking his part in the Gruesome Contest. I worked alone, tediously, precariously, starting in the bow as was proper. Do the hard thing first, right? Climb down the ragged rungs, balancing brush and paint in one hand, cling to the rope with the other hand, both panicked feet. Fight nausea, fight vertigo. Don't look down. Blue water below, silence on deck.

If I fell, a possibility I considered too often and in grisly detail, I'd be on my own in the water. One hundred and ten miles from land. Insulated as they were in the galley, the men wouldn't even hear my shouts.

Paint, paint, paint. Three coats this time. Two of orange primer, one of enamel white. Orange, orange, orange, white, white, white; paint, paint, paint. Every few minutes brace toes on hull and shift rope ladder left, or right. That was the hard part.

Most unnerving of all was when one of the boys, bored with the galley party, hovered drunk over me on the bulwarks, over the two thin strands of my lifeline, talking about knives. Bowie knives and fishing knives, butcher knives, filet knives, stickers and stabbers. Har har har.

Did I say most unnerving? I got ahead of myself. Because the scariest thing was hearing the diesels crank up unexpectedly, as when we were called to the rig. I'd have to scramble up the flimsy rope ladder and safely aboard before the *Condor* lurched off with one terrified painter trailing in her wind and wake. Har har har.

Every night alone in my old room, I overheard the boys' lethal true-tales, their loud unfunny jokes, and now their too-loud bragging of women brought to heel. "That woman sure could suck. Suck like a sucker fish." For background, the sound of knives stropping on the belt that hung from my doorknob. I did not sleep well. My dream guide had deserted.

The men didn't talk to me so much as aim their talk at me. I kept my silence, banked the fire in my eyes. Just survive, painting, counting the days.

Thanks probably to Captain Slicko's instructions, no one crowded me out of the tying and untying of the rig lines, the heaving of the buoy line.

Only one job, by common consent, was barred to me. When we went fishing at the old platforms, occasionally a platform had no bitt to tie up to. One of us must swing over the wheel-wash onto the catwalk. I was at first grateful to be barred from doing it. It's the deck maneuver most heart-catching to watch, a real deckhand killer in hard seas. If there'd been a maritime union in the oilfields, deckhands would have been forbidden to do it, ever, under any conditions.

First the captain backs the boat stern-to the platform's catwalk rail, a precision position that's hard to come by, harder still to hold. The deckhand uses a long gaff hook to snag one of the weathered, rotting lines that trail to the water from the platform overhang fifty or more feet overhead. Then, using the rope for high leverage, he hoists himself up onto the boat's heaving stern rail. From there he swings fifteen or twenty feet across, over the boat's churning wheel-wash, dropping grip on the line at the split second feet clear catwalk rail. If his posture is right and his timing perfect, his feet land firmly on the catwalk and he escapes without a scratch. More often he trips up on the rail, falls and sprawls, snatches any hold at all on the platform to keep from plunging into the sea where the props would thrash him into bloody splinters.

SpaceMan was a master of this Tarzan move. I watched him do it, smoothly. I was maybe only half interested in trying it myself. Too spooky.

But then, during a long fishing day, all the boys swung over to fish from the platform. I was left on the boat, alone, a drudge with her paint, paint, paint.

"Bet you can't do it, Wucy," Danjones called back when he completed his flying leap. I reached for the gaff hook, to try.

Mick vetoed it. "No woman I heard of ever did such of a thing."

I reminded him of my agreement with Captain Slicko. Mick turned away from me, sour, forbidding it still.

Late that night he met me on the back deck. The wind was stiff, chill with a foretaste of winter.

He'd been thinking about me and the platform, he said. He'd decided he'd let me try to swing over there next time if I'd agree to swallow my pride and let him coach me. "There's a trick to it," he said.

Sure I'd let him coach me. I was relieved to hear there was a trick to it.

Two days later we headed for that same platform. Mick put Gitzy on the wheel in his place and joined me at the stern rail for my lesson.

"Don't worry about nothin'," he said. "Just do what I tell you."

But as I braced myself, calling up a kinghell kick of adrenaline for this last big hurdle before I'd be a genuine tried-and-proved deckhand, Mick whistled me over to him. "Lookit down there, Wucy! Sharks!"

And sure enough, a pair of blunt-snouted young tiger sharks circled the platform just below the water's surface.

Sharks.

"Don't you worry about nothin'," Mick said again.

And the show must go on, must it not? I snagged the platform's line with our gaff hook, then hauled myself up to balance on the stern rail. The diesels revved to back the boat in place; the rail surged under my feet. Just below me our props boiled the clear blue Gulf into crystal green bubbles, masking the sharks' circling. Maybe it scared them off.

The space between me and the catwalk was scary enough, seemed a mile away. I reminded myself I'd asked for this, but that didn't keep my knees from locking, my guts from grinding my lunch into a burning gruel in one cramping seizure. This was real. This was death.

I thought I'd been frightened before. I had never been frightened before. Now it was time to swing.

But Mick was tugging at my rope, my one frail hold on life and safety. Reluctantly, I pried my eyes from the platform, saw Mick gesturing that I should . . .

What? Get down?

No, hand him over the rope.

The rope! And stand there on the bucking rail without support? Surely not!

But my captain shook his head impatiently, tugged the rope again.

He must know what he's doing. I let it go.

And then he pushed me over, into the wheelwash.

There was a trick to it, he'd said.

I sucked in air, then water, having time to think I'd best belly-whop to minimize my chances of being sucked into the props, but not having time to carry off that protective maneuver. I went down with arms flailing, and my knee must have glanced off the rusty bumper rail because I had an ugly scrape later. I don't remember the impact. I forgot even the sharks when the sting of seawater rushed up my nose, down my throat.

I heard the diesels die, or should have heard them, because the propwash stopped churning. Gitzy must have cut the power to the props. Remembering the sharks again, I swam, quietly, carefully, for the boat. Sharks are attracted by flailing, I'd heard.

But with its power cut off the *Condor* drew away from me, with the wind. Slowly, inexorably, it drifted out of reach on the blue water. Missing my boat, missing my boat. The last panel of my nightmare was in place.

A jolt of terror shot through me. I might swim, carefully so as not to invite the sharks, back to the abandoned platform. I might somehow climb aboard it. The *Condor* might sail on without me. "We just woke up one morning and she was gone. Must of fell overboard." In two weeks, or two years, another boat might happen by. One hundred and ten miles out in the Gulf, away from the ship channels, traffic was thin, nights were chill, days were empty.

No, it had to be the boat. Forget the sharks, just hope I don't feel the impact if they strike. I struck out swimming hard for the safety of the boat. Through the ring of panic and water in my ears I waited to feel the snap of jaws on my fluttering legs, or to hear the certain death sentence: the roar of power to the *Condor*'s props. No, all I could hear was my own demented unconscious chant: the safety of the boat, the safety of the boat.

Mick stuck out his hand and hauled me aboard himself. I

came up stiff and cold, trailing a wet brown stain from my shorts. He didn't say anything. Neither did I. I just limped across the deck to the house without looking back, without catching Gitzy's eye where he stood at the stern controls. I heard the diesels roar then. I felt . . . ashamed?

I know I cursed my own gullibility and not their guile, their murderous guile. The fear and shock drove all else before, and shame smothered it all, left a hard knot in my throat and an emptiness in the air around my ears.

I went to the head and threw up seawater for a while, then stood with tears brimming in my eyes, surveying the mess I'd made in my shorts. The interior voice was chanting again, only this time what it said was, "Broken. Broken broken broken." Once again I was crouching alone in a cold room wiping up the evidence of my shame, my defeat. Broken, broken, broken.

I started for my own room, but couldn't bring myself to pass the razor strop that hung on my door, the knives that skittered across the galley table with the roll of the boat. No, I would hide. Upstairs, in an unused bunk room, I lay on a bare mattress under two scratchy blankets, shivering.

Broken, broken, broken. Precisely.

I must have slept, for a long time. When I woke I wasn't hungry, or needing to smoke a cigarette, or urinate. I was only empty, and still ashamed. With the same sickening lurch I remembered from waking on the day after my sister's death, the memories rushed back. This had been no nightmare I could wake from. I'd lost it all. There'd been a trick to it.

Pulling the blankets up around my cheeks, I shivered some more. SpaceMan and Danjones had slept through the crime; the only witnesses were its victim, its perpetrators. There would be no point in my telling. What would I tell, anyway? I hadn't yet made a story of it, strung it together with subjects and objects, turning point and climax. Meaning was still missing.

What would Mick tell the others? What could he say?

Wanting a cigarette now, badly, I left the empty bunk with

a blanket wrapped around me and headed downstairs to my room.

"Why here's Miz Wucy!" Mick called out. "We were about to send a scout party for you. I wuz tellin' old SpaceMan here what happened yesterday."

"What did happen yesterday, Mick?" I tried to stir the embers of what had once been my volcano eyes, but I flinched instead, and looked away. "What did happen?"

"Well, you got yourself wet and you learned yourself a lesson."

"What lesson, Mick?"

"Never to trust nobody. When you're out there on that deck, you're out for Number One and nobody else. Isn't that right, Wucy?"

27

I painted out the rest of my hitch on the *Condor,* literally painted the days white, whited them out to suit the hollow in my self. Paint, paint, paint.

I was still painting when we pulled into dock for crew change. Captain Slicko came aboard, carrying a briefcase. He didn't wave at me, or speak. And that was odd, wasn't it?

Alone of his five employees aboard, I was at work. Maybe he'd notice that. The rest were lazing in the galley, honing their stabbers, drunk since Mick's trip to the liquor store. Tomorrow we'd all be going home.

I had no plans to tell anyone, ever, about my going overboard. Certainly not Captain Slicko. Mick and his boys would swear I lied. And I was still ashamed. Here was proof positive that I wasn't capable of blending in, playing dumb. No, I'd move on to the next boat, earn my sixteen days of sea time, get my able bodied seaman's ticket, and then decide whether to go or to stay.

Captain Slicko and Mick were up in the wheelhouse, watching me, it seemed to me, as I cleaned up my painting tools, took out the galley garbage, hosed down the deck. I could almost hear the hiss of my name on the wind. But while I was in the shower Captain Slicko left the boat, still without a word to me, so I guessed I'd never know.

Now that we were at dock our TV had a strong signal again. I was watching an episode of *M.A.S.H.*, relishing the look of all those intelligent Yankee types on one small screen, half wondering why the boys were so eerie quiet that night, hanging out on deck. The galley door opened and a sheriff stepped in, the parish sheriff with his deputy, and Mick behind him.

"Mrs. Gwin? Is there a Mrs. Gwin aboard?"

The things that go through your mind. Through mine: that Mrs. Gwin is the name of my mother, that the sheriff had come to tell me she was injured, or dead. Dear God, please let it not be that. My ears pricked, my skin set up like fresh cement, cold and damp. Here it comes: something bad, real bad.

"I'm Lucy Gwin."

"We want to talk to you a minute."

"What about?"

"About you having drugs aboard this boat."

Then all the puzzling little pieces of the evening fell in place. Slicko's briefcase, the boys' eerie silence, the hiss of my name on the wind. The E.E.O.C. must have notified Slicko that I was planning to continue my case against Byrd Marine.

Possession of drugs and liquor—guns and knives, too, for that matter—aboard a Coast Guard–registered vessel is a federal crime. I'd smoked my last jay the day before, alone, celebrating my one-year anniversary on the boats. So I was clean, but I suspected now that that fact wouldn't matter.

"I seen it," Mick told the sheriff. "In her room, a great big old bag of marijuana, up on the shelf, by the life preservers."

"That's her room right there." Gitzy pointed. I glared at him. He grinned back, a wet pirate-y grin.

Standing off the sheriff at the threshold of my bunk room, I asked if he had a warrant?

"I don't have a warrant, no. But I can get one if you refuse a search. It's more or less of a courtesy search, see, company property and all . . ."

I glared over his shoulder at the boys again.

"Whyn't you boys move on outside," the sheriff said. They did. His deputy, too.

Think fast, think fast there, woman guarding the door, heart in throat. Mick would not have called the law for nothing. The marijuana will be there, on the shelf where he says it is, next to the life preservers. Think fast.

Listen, I said to the sheriff. Here's what I think is happening. I told him about my E.E.O.C. suit, the briefcase, the gleam in Mick's eye. "You can nail me and you know it, just like that. I don't have a lawyer's advice. So you tell me: what should I do?"

"Tell you my own problem," the sheriff said. "All I'm after getting is home in time for the L.S.U. game. First televised game of the season. You make me go for a warrant, it's gotta be a federal warrant. That'll take all night to get, and I'll miss my game.

"So, tell you what. If you let me look, and I do happen to find anything, I'll just kind of like confiscate it, take it out under my jacket, won't never say a word to nobody. Deal?"

Deal.

The dope was easy to find, just where Mick had put it. The sheriff hid it in his jacket, then searched my room noisily, pawing through my locker, my bags, my purse. When he found my pack of rolling papers he took those too. He whispered to me that the bag looked to be "Primo Colombian. Smells *good*. I'd leave you a bit of it, but you better stay clean." He even scratched his home phone number in my notebook, "just in case you want to drop by and burn one at your company's expense."

This was my kind of lawman. Still trembling, I allowed myself a smile.

The sheriff exited my room, shaking his head to Mick, who was waiting just outside. "Captain," the sheriff said, "you just better go and get your eyes examined before you call us all the way out here again."

"Destroyed the evidence!" Mick exploded. "Bitch destroyed the evidence!"

"Don't never trust nobody, Mick," I said when the sheriff was gone.

Captain Slicko had wanted me to be a survivor. He could be proud of me now.

Mick went ashore to call Byrd Marine from the dock's pay phone. I sat back down in the galley, made myself a sweet revenge sandwich, ate it. So tasty. Gitzy joined me for a minute before he went off to bed.

"Wucy, I don't know how you done it, but I got to hand it to you." He shook my hand, then kissed it. I smiled.

Byrd Marine fired me anyway, an hour later. The company carryall was on its way to pick me up, Mick said.

"For nonpossession of marijuana?" I squealed.

But the decision was final. Terminated, again. For the fourth time in one year at sea. My eyes stung, just thinking about it. I slammed off to my bunk to pack, and to cry. Damn them, damn them all.

I heard the boys gather in the galley, laughing loudly, so I could hear them having their last laugh.

And that's when I found the message Betty c. had left: "Big Dicks for the cook." Oh, lady! Oh, please!

28

I wish I could have ended it there, under my own power; kissed off the whole pack of rats and good riddance to the boats, too. Maybe I just hadn't had time to think about it hard enough. Maybe I still cherished some illusion about the prospects of underdogs. Maybe it was simply that, as Aubrey said, the boats get in your blood. Maybe it's that only this book, which was not yet begun, could purge my love for the boats, my grudge against the men.

Because when I left that *Condor*, I left with a hard promise in my heart. They hadn't beaten me yet. I would win my lawsuit, win back my job.

Maybe it was only vanity, or shock.

Whatever it was, it propelled me into one more ugly adventure, that same night.

I slammed into the Byrd Marine van, thinking faster than I could have spoken, feeling more than I could bear. One consoling thought: perhaps this frame-up was a good thing. I already had three types of discrimination and harassment charges on file against Byrd Marine. Now I had a fourth that sealed the deal: retaliation. I decided I'd ask for a boat, the *Heron*, in punitive damages.

"Don't feel bad," the driver, whom I had not yet noticed, said to me. Now I noticed. He was an ugly duck, what little

I could see of him by the green glow of the dash lights. Pitted face, low-lidded eyes, wet lips, slack shoulders. He had, ironic touch, a fat joint tucked behind his ear.

"They fired my best buddy today, too," he told me.

I made a sympathetic noise.

"And that's where they made their big mistake," he said, slowing the van to light his crumpled jay. "Here." He thrust it at me. "Have a toke."

"No thanks, not just now. Maybe later."

"So, as I was sayin', they made a big mistake firin' my main man because now I ain't gonna do what they sent me to do."

"They sent you to do something?"

"But I won't do it. I won't have no part of messin' up a pretty girl like you."

"Mess me up? They sent you to mess me up?"

"I told you don't worry. I wouldn't touch a hair of your pretty head."

The carryall van hurtled through the night, across deserted Pecan Island. Not even a cow in sight. Not another car, not another truck. Night slipped by, whistling on the other side of the windows. I fingered the door lock, as if casually, pulled it up from its locked position calculating the odds of survival if I plunged out of the van, rolled free and into a ditch. Not good. The driver was leaning hard on the accelerator.

"Tell me all about it," I said. "I want to hear this."

"Here." He extended the joint again. "Have a toke. What's your name again?"

"Lucy. No thanks. Really, I don't want to smoke right now. Who told you to mess me up?"

"Look, I wouldn't hurt you if the President of the whole United States told me to, so just relax. I ain't that type of guy. Hey, how old are you? You looked like a young chick when you walked up, but now . . ."

"I'm forty-five years old," I said. "I have a son your age." Lie, lie, slip fingers over the door handle, prepare to bail out when he slows for a curve. Not too many curves on Pecan Island. Dark out there, too. Sign says nineteen more miles to

Intracoastal City. To Guste. Get him to take me to Guste.

"Awww, you're lyin'." The man leaned over at me, inspecting my face. "You ain't no forty-five years old. Are you?"

"Sure am. Going on forty-six."

"Got any pills with you?"

Checking the play in the door handle again, I must have tugged too hard. A roaring inch of space opened up; the wind spun in. "Oops. Slipped."

"Hey, you're scared, ain't you?"

Pulling the door shut again, twice, feeling it bat against the wind. *Bam, bam, bam,* come on door! Whew. There. "Scared? Of you? You look like a nice guy. You wouldn't hurt me."

"Got any pills with you?"

"Pills? No, no, no pills. But I've got some money. You can have it, all of it. Got more money than I know what to do with. Want some? Here, I've got it right here."

"Shit, I could use ten or twenty. No pills, though?"

"No, no pills." Fumbling money out of wallet, ten thirty, fifty, who cares? Let him have it all. "Pills. No. But I know where to get some."

"This time of night?"

"Any time. Right up ahead in Intracoastal City, on my old boat. On the *Pride.*" Now just let the *Pride* be there, let Guste be aboard.

"Sounds good. What'd you say your name was? Lucy?"

"That's right. Lucy. Rhymes with goosey. What's yours?"

The *Pride* was docked at the pipeyard, offloading surplus drill pipe. Guste was up in her wheelhouse with Cupp and even Fred Fatigué, waiting for the crane to finish so they could get back to the oil company dock, to sleep.

I blurted out "Save me!" when I ran up the stairs. Save me, like some silly subtitle on an old movie. Oil Can Harry ties Pretty Pearl to the railroad tracks. She cries, "Save me!"

Guste gathered my shivering shoulders into his arms and tried to calm me. I stuttered out what was happening: that van out in the lot, the one with the interior lights on, had a murderer in it. My bags, too.

My friends formed a posse, walked with me back to the van to collect my bags. "Jus' run yo'seff on home now, you," Guste told the boy.

The boy looked at me as though I'd betrayed him. And then he drove away. Ten minutes later I was asleep on the galley bench, on the *Pride*.

Guste shook me awake the next morning, wanting to hear the whole story of the night before. I told him about the stabbers, the Jacob's ladder, the dope, the sheriff, the terrifying ride home. I left out the going overboard. That would have been too much. It was too much already.

But Guste offered up another round of condolences. "Tchoo," he said, and "awww."

I was uneasy with his sympathy, with my one horror story too many. Terrible things happen to terrible people, I kept thinking. I tried not to cry.

Guste had some news of his own. The deal with Petrolco had gone through. A week from now he'd be taking over their *Capricorn*. He'd already arranged with them to take me along as deckhand. "We tear dem a new asshole, us," he said.

"We sure will," I agreed, all teary-eyed and grinning like a fool.

How many times will a heart leap for a happy ending? Would I never learn?

After breakfast, Guste walked me out to the carryall that was taking Cupp and the eastbound crew home for their leave. I had to ask Guste one more time: was he absolutely sure he'd go through with it?

"Sho' will do," Guste said. "Now give yo' ol' cap'n a kiss fo' dat."

I kissed his cheek, the usual daughterly peck with an extra oomph for joy.

"Naw, naw, not like dat dair. Give some real kissin' dis time."

And then he grasped my cheeks, and kissed my lips, hard. Grasping me harder, he stuck his hard wet tongue in my mouth.

It's been just about two years now since the boats started leaving without me. I still leap to catch them in my dreams and wake up holding my breath, hearing my heart pound in my ears. Sometimes, sitting there alone in the dark, I close my eyes and rock on my bed, arrythmically, like a boat on the water.

Acknowledgments

As of this writing, more than twenty women work as deckhands on supply vessels in the Gulf Coast offshore oilfields. Godspeed, women.

I want to thank Verna Gwin and Sue Dawson for saving the letters from which the first draft of this book was made.

The writing of this book was subsidized, albeit unwillingly, by a settlement from the boat company identified here as "Byrd Marine." My thanks to Jackie Marv, Eddie Burks, Nancy Meierding, Michael Smith, David Lewandowski, Bayou Lafourche Legal Services, Inc., and the Equal Employment Opportunity Commission for their help in obtaining that subsidy.